PENGUIN BOOKS

THE GREEN CONSUMER

JOEL MAKOWER is a Washington, DC–based writer and lecturer on consumer and business environmental topics, and editor of *The Green Consumer Letter* and *The Green Business Letter*. His weekly syndicated column, "The Green Consumer," appears in newspapers throughout the United States. He is author or coauthor of more than a dozen books, including *The Green Consumer Supermarket Guide*; *Going Green: A Kid's Handbook to Saving the Planet; The Green Commuter: Driving (and Not Driving) for a Healthier, Cleaner World; 50 Simple Things Your Business Can Do to Save the Earth;* and *The E-Factor: The Bottom-Line Approach to Environmentally Responsible Business.*

JOHN ELKINGTON is one of Europe's leading authorities on the role of industry in sustainable development. He runs an independent consultancy whose clients have included British Petroleum, the Nature Conservancy Council, and the United Nations Environment Programme. Having written numerous books and reports, he is also the co-author, with Tom Burke, of *The Green Capitalists.*

JULIA HAILES has worked in advertising and TV production. In 1987, she helped to set up SustainAbility, "the green growth company," with John Elkington and Tom Burke. The company aims to promote environmentally sustainable economic growth.

REVISED EDITION

THE GREEN CONSUMER

JOEL MAKOWER

WITH JOHN ELKINGTON AND JULIA HAILES

A Tilden Press Book

PENGUIN BOOKS

PENGUIN BOOKS
Published by the Penguin Group
Penguin Books USA Inc., 375 Hudson Street,
New York, New York 10014, U.S.A.
Penguin Books Ltd, 27 Wrights Lane,
London W8 5TZ, England
Penguin Books Australia Ltd, Ringwood,
Victoria, Australia
Penguin Books Canada Ltd, 10 Alcorn Avenue,
Toronto, Ontario, Canada M4V 3B2
Penguin Books (N.Z.) Ltd, 182–190 Wairau Road,
Auckland 10, New Zealand

Penguin Books Ltd, Registered Offices:
Harmondsworth, Middlesex, England

First published in Great Britain by Victor Gollancz Ltd. 1988
Updated edition published in the United States of America in Penguin Books 1990
This revised edition published in Penguin Books 1993

5 7 9 10 8 6 4

LIBRARY OF CONGRESS CATALOGING IN PUBLICATION DATA
Makower, Joel
The green consumer (revised edition) / Joel Makower, with John Elkington and Julia Hailes
p. cm.
Includes bibliographical references
ISBN 0 14 01.7711 6
1. Consumption (Economics)—United States. 2. Environmental protection—United States. I. Elkington, John. II. Hailes, Julia.
III. Title
HC110.C6E44 1993
363.7'057—dc20 89–48639

Printed in the United States of America
Set in Janson and Futura Condensed
Typesetting by MacSachel Imagesetting, Wheaton, Maryland

About This Book

We have divided *The Green Consumer* into three parts. Part I offers the broad view of Green Consumerism, the state of the environment, and how your everyday purchases can affect the earth's resources. Part II features a comprehensive guide to what products to buy—and not to buy—including brand names and addresses. Part III is for those who want to learn more or become more involved in environmental issues, including names and addresses of dozens of environmental organizations.

There are two principal ways this book can be of use. One, of course, is through the listings of green products and companies. Understand, however, that a product's inclusion in this book does not constitute an endorsement of this product, nor does it ensure a product's quality and efficacy. The second and equally important use of this book is to obtain a better understanding that will help you to judge *all* products for their potential environmental impact. By learning to include environmental concerns along with price, quality, nutrition, and convenience, you will be very well equipped to meet the demanding task of being a Green Consumer in the 1990s and beyond.

When contacting any of the companies in this book, be aware that each has its own policies and practices. Some sell directly to consumers, accepting phone orders and credit card charges. Others accept orders by mail, accompanied with payment. (Note that prices listed in this book are suggested retail prices and do not include shipping or sales taxes.) Still other companies will refer you to a local retailer that carries their products. Some companies offer catalogs and brochures of their products, while others have no such information; although most of this literature is free, there may be a charge for some, or a requirement to send a stamped, self-addressed envelope. Moreover, all of these policies are subject to change. As this book went to press, for example, several companies that had not previously sold directly to consumers reported that they were considering doing so. Others were preparing—or thinking of preparing—brochures or catalogs for the first time.

We hope you find the information contained in these pages to be valuable. We welcome your ideas, suggestions, and comments, which we will incorporate in future editions of *The Green Consumer*. Please send them to The Green Consumer, c/o Tilden Press, Inc., 1526 Connecticut Ave. NW, Washington, DC 20036.

Authors' Note

The authors and publishers of the American edition of this work have prepared this work and set forth the views contained in it in good faith.

The omission of any particular product, company, or any other organization from this work implies neither censure nor recommendation. Neither the authors nor the publishers of this edition of the work, however, warrant effectiveness or performance of any product described in this work, and the reader and consumer must exercise his or her own judgment when determining what criteria to apply when judging whether he or she should purchase any particular product from any of those companies or other organizations. Any reader requiring further information concerning any particular product should write directly to the supplier or manufacturer of that product.

The authors and publishers are willing to consider any matter which is held to be inaccurate and, upon satisfactory factual and documentary confirmation of the correction, will use all reasonable endeavors to amend the text for the next reprint or edition.

While every care is used and good faith is exercised in assembling and publishing the material in this book, no warranty as to the properties or safety of any product is imputed by the inclusion of any such product in this work, and consumers should address themselves to the manufacturers of any product in respect of which they may wish to be informed as to its qualities.

CONTENTS

Acknowledgments x
Foreword by Ben Cohen xi

Part I: The Green Consumer

Introduction 3
The Three R's 10
Seven Environmental Problems You Can Do Something About 14
The Problem of Packaging 33
The Recycling Solution 44
Eco-Labeling 51

Part II: Shopping for a Better Environment

Automobiles 55
The Elusive Electric Car 61
Buying Green 64
Driving Green 68
A Fuel's Paradise 74
Disposing of Hazardous Substances 79
The Green Commuter 85

Clothing & Personal Care Products 89
The Fabrics of Our Lives 90
The Diaper Dilemma 98
Earth-Friendly Cosmetics 102
Cruelty-Free Products 108

Food & Groceries **115**
How Safe Is Our Food? 118
Organic Foods 123
Along the Shelves 126
Paper Products 153
Greener Cleaners 158
Products for the Ages 166
Natural Cleaners 168

Garden & Pet Supplies **175**
Problems with Pesticides 176
Natural Remedies for Pests and Diseases 182
Healthy Lawns, Healthy People 185
Composting 190
How Trees Save the Earth 192
Dealing with Pets' Pests 195

Home Energy & Furnishings **203**
Energy-Efficient Appliances 204
Products to Save Energy 228
Green Lights 233
Rechargeable Batteries 241
Saving Water at Home 244
Ecological House Paint 247
Earth-Friendly Furniture 253
Good Wood 258

Responsible Travel **261**
Eco-Tour Organizations 266

Toys & Gifts **273**
A Grab-Bag of Eco-Toys 275
Green Gifts for Everyone 277
Alternative Trade Organizations 282

Part III: How to Get Involved

How to Get Involved 287
Taking On Local Polluters 288
The Power of Consumer Boycotts 294
Investing in the Environment 298
Making Your Business Green 302
Environmental Organizations 306

Index 332

Acknowledgments

A great many individuals and organizations helped in this effort, and they deserve recognition and thanks. First and foremost, we would like to thank the dozens of individuals at environmental groups, government agencies, manufacturers, and retailers, as well as countless other experts with whom we consulted. In addition, we must acknowledge the staff—past and present—of *The Green Consumer Letter*, who researched and wrote newsletter stories that have been excerpted or adapted in this book: Alex Friend, Ilyanna Kreske, Norman Meres, Anna Mulrine, Jamie Queoff, and Nancy Tienvieri.

Thanks also to several individuals whose assistance went beyond the call of duty: John Morrill and Glee Murray, American Council for an Energy-Efficient Economy; Richard Dennison, Environmental Defense Fund; Liz Cook, Friends of the Earth; Carla Garrison, The Ecotourism Society; Alice Tepper Marlin, Council on Economic Priorities; Pamela Dorman and Janet Kraybill, Penguin Books; John Javna, The Earthworks Group; Linda Malone, Real Goods Trading Corp.; and graphic designer Mary Ann W. Bruce.

Foreword

by Ben Cohen, Chairperson
Ben & Jerry's Homemade

When Ben & Jerry's first started, I used to be a hot ticket on the Rotary Club speaker circuit. The assignment was to drone on a bit after lunch in order to help people digest their meal. Ben & Jerry's was a real small company at the time—we were a homemade ice cream parlor—and we used to do things for the community, like show free movies during the summer on the wall outside our gas station, and sponsor community celebrations, and give away free ice cream on our anniversary. Things like that.

I'd be talking to the Rotary Club about these things and at the end of the talk somebody would kind of lift up his head and say, "Well, you know, those things you're doing for the community—you're just doing them because it's good for business, right?" And I responded, "Well, I don't know, but our reason is that we genuinely believe that business has a responsibility to give back to the community, and we're doing it out of altruism."

That was my old answer. My new answer is, "Yeah, it is good for business. And if you're smart, you'll jump on the bandwagon."

I have been thinking a lot about the influencers of business. One of those influencers is capital—that is, the money that people invest in businesses by buying stocks. Many people are starting to invest their money only in companies whose values they agree with. Businesses need this investment capital in order to survive and prosper. So, as more and more investment capital has these kinds of strings attached to it, companies are starting to be influenced to go along with investors' values. Another influencer is the value of employees. Employees run and operate businesses. So, it makes sense to start educating business school students in the art and science of operating businesses in such a way that they make a profit and proactively support our communities—local, national, and global—at the same time. A third influencer is sales. Businesses need sales to exist, and if companies find that customers will support them more if they adopt a particular social stance, they are going to move in that direction.

That's all fine and good. But consumers need the tools with which to vote with their dollars and decide which businesses to support and which businesses have values that they agree with and which businesses don't. That's where being a Green Consumer comes in.

The beauty of it is that being a Green Consumer is a really easy thing to do. All you need is some education. You don't have to go out of your way in order to influence companies through your purchasing behavior. When you go into the supermarket, or any store for that matter, you usually have quite a few choices of brands to buy if you're looking for a particular product. In many cases we're talking about products between which there really isn't much difference. You simply choose which packaging you want or which color you want or which one has a slightly better price or a name that rings a bell in your head. Now, with this book, you can add another factor: Is this product and this company environmentally responsible?

Companies that want to survive into the twenty-first century know that they must consider values. At Ben & Jerry's, we operate as a values-led business, a term coined by Anita Roddick, founder of The Body Shop. We tend to pursue activities based on a particular set of values. So if our value is to create a world that spends less on the military, we will find a way to integrate that message into our day-to-day business activities and influence our government in that direction. If our value is to try to create less economic disparity in our society, we will support organizations that are working in that direction through our purchasing decisions, our investment decisions, and our internal salary structure. If our value is to try to help the environment, we will support that through internal company recycling programs, through externally trying to educate people about recycling, and by working with companies like Community Products, whose purpose is to prove that rain forests can be more profitable as living rain forests than as deforested land.

I think the real problem is that we need to redefine the bottom line for business success. The bottom line has to be in two parts. When we measure our success as a business at the end of the year, we must look at how much money is left over and how much of a contribution we made to the community—whether it's the local community or the nation or the world. We have to factor those two things together to determine how well we're doing.

As Green Consumers, you have the right and responsibility to vote—with your dollars—on how well businesses are doing these things: how successfully they are addressing the issues you believe are most important to your life, and to the life of our planet.

As the saying goes, vote early, and vote often.

Part I

THE GREEN CONSUMER

INTRODUCTION

You probably don't realize it, but every week you make dozens of decisions that directly affect the environment of the planet Earth. At work, at home, and at play, whether shopping for life's basic necessities or its most indulgent luxuries, the choices you make are a never-ending series of votes for or against the environment.

Buy a burger, fries, and a soda and you are probably worsening the already critical landfill crisis. Take your car in for repairs and you may be contributing to the gradual warming of the earth and increasing your chances of getting skin cancer. Do your laundry and you may be fouling America's lakes and rivers, perhaps your own drinking water. Discard your trash and you may be polluting the air, water, and soil, and helping to deplete the earth's natural resources.

But the products and services you buy need not be so destructive to the environment. By choosing carefully, you can have a positive impact on the environment without significantly compromising your way of life. That's what being a Green Consumer is all about.

It wasn't very long ago that being a Green Consumer was a contradiction in terms. To truly care for the environment, it was said, you had to drastically reduce your purchases of everything—food, clothing, appliances, and other "lifestyle" items—to a bare minimum. That approach simply doesn't work in our increasingly convenience- and consumption-oriented society. No one wants to go back to a less-comfortable, less-convenient way of life.

And yet in the past few years consumers in many nations have demonstrated their environmental concern through the products they buy—and don't buy. While the supermarket, hardware store, and department store of today doesn't look a lot different from that of a few years ago, some dramatic changes have taken place, both on the shelves and behind the scenes. Due largely to consumer pressure, companies are "greening" their products and their operations, reducing packaging, increasing recycling, and otherwise delivering their goods with less impact on the environment.

But these changes are only a small beginning of what is needed to change the impact of our purchases on the environment. Many of our most commonly purchased products are still made by inefficient, wasteful, polluting companies. Many are overpackaged, or manufactured in ways that are destructive to the air, water, land, or various creatures, from microorganisms to human beings. It is up to Green Consumers to pressure these companies to change their products and processes.

This book will show you how to do this, in part by directing your purchases to environmentally responsible products and companies. It is intended to help you sort through the often confusing world of Green Consumerism to make good, green choices.

Why Shop Green?

Why should you do this? For starters, as you'll see, some of the products we buy contribute to environmental problems. You may be surprised at the ways this can happen. For example, the manufacture of a paper towel or napkin can contribute dangerous pollutants downstream from a paper mill. The use of some aerosol products contributes to urban smog. Many products are packaged excessively in materials that are neither recycled nor recyclable. Some products are made by companies that have poor environmental records. Buying products from companies that pollute supports their lack of concern for the environment. Bigger products like cars and appliances can have major impacts on a variety of environmental problems. Even the gifts you give and the kind of vacations you take can harm—or help—the Earth.

Make no mistake: We're not suggesting we can simply shop our way to environmental health. Part of being a Green Consumer is learning when *not* to buy—when more is not necessarily better. By making the right choices, you can help to minimize the pollution and waste created by many "un-green" products and companies.

How to Be a Green Consumer

When it comes to evaluating and choosing green products—those that cause minimal impact on the environment—here are ten basic guidelines you should consider:

- Look for products packaged in readily recyclable materials such as cardboard, aluminum, or glass.
- Don't buy products that are excessively packaged or wrapped.
- Look for products in reusable containers or for which concentrated refills may be purchased.
- Look for products made from the highest content possible of recycled paper, aluminum, glass, plastic, and other materials.
- Choose simple products containing the least amount of bleaches, dyes, and fragrances.
- Look beyond the products to the companies that make them. Support those with good environmental records.
- Don't confuse "green" with "healthy." Not everything packaged in recyclable packaging is necessarily good for you or the environment.
- Remember that there are few perfectly green products.
- Keep in mind that doing something is better than doing nothing. Every purchase you make counts.
- It doesn't matter much whether you carry your purchases home in a paper or plastic bag. What's important is that you reuse or recycle whatever bag you get.

Keeping Things in Perspective

It's important to keep things in perspective. Green Consumerism is not the ultimate solution to the world's environmental problems. The problems are simply too many and too complex to be resolved merely by smart shopping. For example, for all the debate over paper versus plastic shopping bags, or glass versus aluminum containers, the environmental differences between these choices are obliterated by the energy and pollution impact of driving just one mile and back to the supermarket in a poorly tuned, gas-guzzling car with underinflated tires. In other words, the mere act of getting to and from the store can have a greater environmental impact than any of the purchases you make on that shopping trip.

That doesn't mean, however, that being a Green Consumer is

not important. Once you begin to incorporate the environment into your everyday shopping trips, it won't be a major step to then incorporate it into other aspects of your life—such as voting, investing, and educating your kids. And so in that light, Green Consumerism takes on a new importance: If we can't sell the environment at Safeway and 7-Eleven, we probably won't be able to sell it in the schools, the statehouse, and the Senate.

That makes your everyday shopping trips one key to saving the Earth.

Who's Shopping Green?

It seems that nearly every month there's another study that shows that Green Consumerism is a driving force in the American marketplace. One study shows that 89 percent of Americans would pick Product A over Product B if Product A was kinder and gentler to the planet. A Gallup Poll found that about three out of four Americans consider themselves "environmentalists." Another tells us that 82 percent of us are voluntarily recycling some materials. Still another survey revealed that three-fourths of all Americans are willing to pay a bit more for environmentally responsible products. A Louis Harris poll revealed that "a clean environment" ranked second, just behind "a happy family life" among American's priorities. ("A satisfying sex life" ranked only in fifth place.) Clearly, the eco-number-crunchers are working overtime.

Don't accept these polls and surveys at face value. While there certainly is increased interest in the environmental impact of one's purchases, there is a very big step from environmental *concern* to environmental *consumerism*. For example, that eight out of ten Americans claim they are recycling "some materials" is a bit suspect in a nation that recycles only about a tenth of all its trash. The fact is, when it comes to the environment, people are more likely to give the "right" answer than the "real" answer. But even if the polls are off by a factor of three or four or five, we're still talking about a significant chunk of the marketplace. And that translates into a great deal of potential consumer clout.

The Price Is Right

One of the most infuriating questions asked by some of the polls and surveys is how much extra Americans would be willing to pay for green products. In most cases, the majority of respondents claim that they would be willing to pay between 5 percent and 15 percent extra for environmentally responsible products. But the question is itself misleading: It furthers the myth that green products are by their very nature more expensive than their "ungreen" counterparts.

The fact is this: Being a Green Consumer can be a money-saving proposition. And with good reason. At its core, Green Consumerism has to do with cutting waste and making the most of one's resources. You can't help but get more for your money in the process.

This notion isn't based on mere speculation. In 1990, two supermarket industry consultants asked Information Resources Inc., a market-survey firm, to track several green products in a major West Coast supermarket chain. The products included those made from or packaged in recycled material, photo- and biodegradable products, those whose "use does not pollute water or air," and products "made using a cleaner process." The researchers were surprised by their own findings. "Consumers weren't in fact paying more for these brands," they told *Progressive Grocer*, a trade magazine. "On average they actually paid between 8 percent and 22 percent less for the green brands contrasted to the average prices paid in the category." But this isn't universally true: Some greener products do cost considerably more than their less-green counterparts.

So, being a Green Consumer can be both ecological and economical.

The Many Shades of Green

One of the first things you find when you take a look at the world of Green Consumerism is that there are few *perfectly* green products. In fact, there are many shades of green. There are green products packaged in un-green packaging, green packaging enclosing un-green products, and green products made by decidedly un-green companies. To make matters worse, you can buy green products from un-green supermarkets, and un-green products at green supermarkets.

It's also important to point out that compiling a list of products that are safe for the environment is simple: There are none. That's right: Nothing is safe for the environment. Its mere existence means it's had some environmental impact. The goal, then, is to choose products that have a minimal impact on the environment. (One of the first things you can do immediately is to discount any product whose manufacturer claims it to be "safe for the environment.")

Confused? That's understandable. But don't let it discourage you. All it takes to begin making green choices are a few ground rules—and this book.

WHAT MAKES A PRODUCT GREEN?

There are few standards or laws that determine what makes a product environmentally responsible. But there are some basic criteria we use when we look at a product's greenness. For example, a truly green product is one that:

- is not dangerous to the health of people or animals.
- causes minimal damage to the environment during its manufacture, use, and disposal.
- does not consume a disproportionate amount of energy or other resources during its manufacture, use, or disposal.
- does not cause unnecessary waste, due either to excessive packaging or to a short useful life.
- does not cause unnecessary cruelty to animals.
- does not use materials derived from threatened species.
- does not cost any more than its "ungreen" counterpart.

That's a very tall order, to be sure. And few, if any, products contained in this book meet all of these criteria. But there is no question that due to increased consumer interest, a growing number of products are coming closer to hitting the mark.

Don't be discouraged by this lack of perfection. There are still green choices to be made in nearly every supermarket aisle. Many familiar products may be deemed "green" simply because they have minimal packaging, or because they are packaged in recyclable containers. The point is, there are plenty of green products to choose from, if you know what you're looking for.

Good, Green Choices

It is also very important to make the distinction between green products and "good, green choices." The former are products that make specific environmental claims—about being recycled, non-toxic, degradable, or whatever. These kinds of products are still relatively scarce in supermarkets and other retailers, and not all of them live up to their claims.

The other type of products—good, green choices—are everywhere. These products are environmentally preferable even though their manufacturers don't necessarily make specific green claims. For example, choosing ketchup in a glass bottle instead of a plastic one—assuming you can (and will) recycle the glass in your community and that the plastic bottle isn't recyclable—is a good green choice. So, too, is buying products in bulk to avoid excess packaging, or choosing a product with fewer layers of packaging. All of these are wise choices for Green Consumers, and can be just as important in voicing your environmental concern as is buying products that make specific claims.

The E-Factor

You already have a set of standards and preferences you bring with you when you walk up and down the supermarket aisles. You think about price, of course, and quality. You consider brand reputation and availability. And you think a lot about your own personal taste.

Each of those factors plays a role differently with every purchase. When choosing a brand of chicken soup, for example, you might say to yourself, "They all taste the same to me, so I'm going to pick the cheapest one." In that case, price is the determining factor. Or you might say, "I've always found Smith's to be a quality brand, so no matter what it costs, I'll buy their chicken soup." And so, brand name and reputation become the deciding factors. We make these decisions—consciously or unconsciously—with nearly everything we buy. These are the tools with which we shop.

Being a Green Consumer means adding another tool: the E-Factor, the environment. So, in choosing your brand of chicken soup you might decide, "I'll choose this one because it has fewer layers of packaging." Or, "This company has a good environmental record, so I'll choose their brand." The E-Factor may not play a role in every purchase, but neither does price, brand name, or quality. But you may be surprised to find how in many purchases the E-Factor does play a role.

The more you consider the E-Factor, the more you will find it a useful way to look at products—not just in the supermarket, but also in the drug store, the hardware store, the appliance center, the garden center, the automobile showroom—just about anywhere else you shop.

THE THREE R'S

To help us remember what to do when shopping, the Green Consumer movement has its own set of "Three R's": reduce, reuse, and recycle. Each of them plays a key role in our attempts to minimize the environmental problems caused by our purchases and lifestyles. Keep in mind that this is a hierarchy: they are listed in descending order of preference.

1. Reduce

This is where the power of green is at its strongest. By avoiding buying wasteful and polluting products, you can make a powerful statement. As we have learned in other cases—most notably in the tuna industry's decision to go "dolphin-safe," and in McDonald's decision to switch from polystyrene to paper—a few Green Consumers can send a loud and clear message to companies to change the way they do business.

What kinds of purchases should you avoid? Here are a few:

- products wrapped in many layers of packaging
- products packaged in unrecycled or unrecyclable materials
- single-use and other products that have a short life before they must be thrown away
- products that are not energy efficient
- products made by companies with poor environmental records
- products purchased from retailers that have poor environmental records
- products with misleading claims about their "greenness"

Enough?

The August 1, 1955, issue of *Life* offered a two-page piece on "Throwaway Living." With a photo of a happy family tossing dozens of disposables into the air, it celebrated these products' ability to "cut down on household chores."

In hindsight, says Alan Durning, we might have been better off applying a bit more elbow grease.

Durning is author of *How Much Is Enough? The Consumer Society and the Future of the Earth*, a brave new book destined to be studied and debated for years. In 200 powerful pages, Durning manages to credit the American dream—and its counterpart in other countries making up "the affluent fifth of humanity"—with causing the lion's share of environmental ills, from ozone destruction to global warming to disappearing species.

"Over a few short generations, we have become car drivers, television watchers, junk-food eaters, mall shoppers, and throwaway buyers. The rest of the world, meanwhile, is watching us on television and now aspires to our consumer life-style."

The unstated goal, he says, "is to lift everyone into the consumer class, but the global environment cannot support even 1 billion of us living the American dream, let alone 5 billion."

Of course, it's not just the rich that contribute to these ills. The poorest fifth also wreak havoc on the Earth, by devastating their land in an attempt to eke out survival. But in seeking a better life for them the question remains: "How much is enough?"

Durning doesn't advocate that we go back to hunting and gathering. Rather, he offers suggestions for incremental change—encouraging workers to trade wage increases for more leisure time, or ending consumption-inducing taxes and subsidies. In the end, the book raises far more questions than it answers, but they are good questions that need to be asked.

With hard work and a little luck, the answers will come.

The book is $8.95 in bookstores or $11.95 postpaid from Worldwatch Institute, 1776 Massachusetts Ave. NW, Washington, DC 20036; 202-452-1999.

As we said before, few products are perfect. Your level of refusal to purchase some of these products will likely be influenced by the available alternatives. If a product you feel you need is available only in one form, and it is an environmentally undesirable one, you may choose to buy it anyway. But you don't have to accept this as "the best you can get." Consider writing the manufacturer and asking them to consider changing the product's packaging or contents to make it more environmentally responsible. Your letter will have more impact, however, if you have chosen not to purchase the product, and tell the manufacturer that.

"Reduce" is also known as "precycling," that is, eliminating wasteful products *before* you purchase them, rather than having to reuse, recycle, or otherwise dispose of them later on.

2. Reuse

Things that may only be used once before being thrown away are an inefficient use of our precious resources. It would be ideal if the products you do buy had the longest life possible. So, it is important to buy products that can be reused over and over. Consider batteries, for example. Why purchase batteries that must be thrown away—filling landfills with a melange of hazardous chemicals—when you can buy ones that are rechargeable hundreds of times? Why buy something that will have a short life when you can buy something that will last and last?

Some reusable things may cost a bit more to buy—rechargeable batteries, for example, are considerably more expensive than disposable ones—but over time, most of these products can more than pay for themselves. For example, a battery charger and four "AA" batteries sell for about $15, compared to four disposable "AA" batteries, which retail for around $3. But if you recharge the batteries only five times, you'll save enough money to recover the cost of the equipment, plus the electricity used for recharging. After that, you're ahead of the game!

Another aspect of "reuse" is to look for products made from or packaged in recycled material. By doing this you are supporting the reuse of resources. The greater the content of recycled material, the better. Some products or packages state specifically: "Made of 100% recycled content." Lacking such statements, it's difficult to tell the exact amount of recycled content.

3. Recycle

If you have refused and reused as much as possible, a high percentage of your leftover trash should be recyclable. And it is extremely important that you make sure to recycle what's left, so that we can get the most use out of all the resources we use and the materials and products we manufacture.

Reuse, Refurbish, Repair

Another aspect of "reuse" has to do with the stuff we throw out every day that can be made into something else, used by someone else, refurbished, or repaired. In fact, this idea has inspired a new breed of second-hand shop: the reuse center.

There are several types of centers. The East Bay Depot in Oakland, CA, is primarily a place where teachers and artists come to rummage through stacks of donated paper, wood, bolts of cloth, and other materials. Another example is Urban Ore Inc., in nearby Berkeley, which accepts furniture, rugs, housewares, books, and office equipment. The company also runs a building materials exchange for doors, windows, sinks, tubs, toilets, lumber, bricks, glass, and hardware. In 1991 it did nearly $1 million in sales.

Another type of dealer is Robert Langeland's Wastebusters on New York City's Staten Island. A former construction worker and contractor, Langeland was always finding nearly new, serviceable doors, sinks, windows, and hardware in dumpsters at construction sites. He now removes furniture and fixtures from office buildings and apartments that are being renovated, and sells them out of a warehouse at cut-rate prices. Unlike high-end dealers of architectural elements that more closely resemble antique stores, Wastebusters sells practical, low-end items, including partitions, carpeting, and floor tiles. It's essentially a salvage business.

Then there's Olmstead County, MN, home of the city of Rochester. The county's hazardous-waste collection site began a product exchange. "We inspect every container that comes in, and if they're good, we put them on shelves and people and businesses can come in six days a week and pick them up," explains Jack Stansfield, the county's waste reduction coordinator. "Our savings is we don't have to dispose of them as hazardous waste, and people don't have to buy them new." Some of the cans sitting on the shelves are weed killers, wood preservatives, varnishes, and paint thinner. People have painted warehouses and farm buildings with free paint from the exchange.

A number of cities have compiled local reuse directories. Perhaps the best example is *Use it Again, Seattle!: A Money-Saving Guide to Repairing, Reusing, and Renting Goods in Seattle*, a 52-page booklet that lists repair shops, second-hand stores, and rental companies. A big attraction are coupons for many stores listed in the booklet, offering discounts or free services. For more information, contact the Seattle Solid Waste Utility, 710 2nd Ave., Ste. 505, Seattle, WA 98104; 206-684-4684.

More on recycling in a moment. For now, keep in mind that for all the emphasis in our society about recycling, in the "Three-Rs" hierarchy recycling is a third choice—after avoiding wasteful purchases altogether and reusing things as much as possible.

SEVEN ENVIRONMENTAL PROBLEMS YOU CAN DO SOMETHING ABOUT

You've probably been hearing for years about "the environment." But how much do you really understand the earth's environmental problems? A basic working knowledge of the biggest environmental concerns is the first step to being a Green Consumer. By understanding the problems, you will best be able to understand how your consumer purchases may be contributing to them.

While a technically thorough explanation of environmental causes and effects could fill an entire book, below, in no particular order of importance, is a summary of the biggest environmental problems of the 1990s and beyond.

1. Acid Rain

We have usually considered rain to be a cleansing experience. A good rainfall is thought to "clear the air" and give everything a nice fresh smell. But in certain parts of the United States as well as in other parts of the world, air pollutants mix with rainfall to pollute rivers and lakes and kill trees. Acid rain has been known to peel the paint off cars and is suspected of damaging the Canadian maple tree population so severely that the maple syrup industry there may be defunct by the turn of the century. Worst of all, acid rain is also known to represent a major health hazard to people.

The two biggest pollutants that contribute to acid rain are sulfur dioxide, which comes primarily from electric utilities' burning of high-sulfur coal, and nitrogen oxides, which also result from utility combustion, as well as the engines of cars and trucks. Once in the air, sulfur dioxide and nitrogen oxides combine with other airborne chemicals and water to form sulfuric acid and nitric acid, and undergo further chemical reactions to become sulfates and nitrates. When these chemicals mix with rain, sleet, snow, or hail, they fall to earth, where they can wreak havoc on just about everything. Even during dry weather, acidic particles or gases can form on the ground, where they mix with soil or are taken up directly by plants.

One of the biggest problems of acid rain is the acidification it causes in lakes and streams, killing fish and other aquatic life. In the

Adirondack Mountains of New York, numerous clear lakes believed to have supported fish life at one time have been found to be devoid of fish. The acidity levels of these lakes is considerably higher than those of nearby lakes that still contain fish. In addition, acid rain affects drinking water by causing several toxic materials to leach into drinking water: aluminum (which has been connected with Alzheimer's and Parkinson's diseases and central nervous system disorders), asbestos (a known cause of lung cancer and other respiratory diseases), cadmium (associated with kidney damage), and lead (associated with brain damage in children and increased risk of hypertension and heart disease in adult males). These toxins are picked up from corroded water pipes, the solder that holds the pipes together, and from whatever soil the water passes through before reaching reservoirs or seeping into underground aquifers—including city dumps and toxic landfills.

Most important, acid rain is being increasingly seen as a threat to human health. For one thing, the chemicals that cause acid rain can also harm the respiratory system. Most vulnerable are the very young and the elderly, particularly those already affected by asthma and bronchitis. Also at risk are pregnant women and people with heart disease. According to Dr. Philip J. Landrigan, a professor of community medicine and pediatrics at the Mount Sinai School of Medicine, acid rain may be the third largest cause of lung disease, after smoking and passive smoking. A congressional study blamed acid rain for contributing to some 50,000 premature deaths a year in the United States and Canada.

What You Can Do. There are limits to what individuals can do to control acid rain. However, there are at least two major solutions:

- **Conserve energy.** Much of the problem results from industrial emissions, especially electricity-generating utilities that use high-sulfur coal. Reducing the need to build additional power plants is one effective means of easing the problem. At home, energy-wasting appliances are a major and needless drain on our resources. Switching to energy-efficient appliances not only saves money over the long term but also drastically reduces energy utility needs. According to the American Council for an Energy-Efficient Economy, if every household in the U.S. had the most energy-efficient refrigerators currently available, the electricity savings would eliminate the need for about ten large power plants.

• **Drive a fuel-efficient car.** Nitrogen oxides in automobile and truck exhaust are a principal ingredient in the creation of acid rain. There is little argument among the experts that driving efficient automobiles can lead to a significant reduction of that problem. See the "Automobile" chapter for additional information.

2. Global Warming

The world is warming, say the experts, and it is directly the result of the gases and pollutants we spew into the atmosphere. No one knows exactly how much the earth may be heating up, how fast this is happening, or what the effects will be. But this much is known: Climatic zones are shifting, glaciers are melting, and the levels of our oceans are rising. By most estimates, as warming continues, forests will die, coastal areas will flood, the world's agricultural areas will wither, and there will be great economic upheaval.

Why are these things happening? Since the beginning of time, the sun has been beating down on the earth, providing warmth and light. Once it reaches the earth, some of that heat bounces back into space while the rest is trapped by a layer of gases that surround the Earth. This is known as the "greenhouse effect" because it mimics what goes on inside a greenhouse: Sunlight shines through the glass, which traps the heat from radiating back outside. So, even on a cold day, it can be very warm inside a greenhouse.

Without the greenhouse effect, the world would be too cold to be livable by humans. But due to human activity, we have increased the amount of gases in the atmosphere, which has resulted in the Earth's "greenhouse" trapping more heat. The result is a gradual warming of the Earth.

Global warming has serious implications for future generations. With the use of computer models, scientists have predicted that average global temperatures will increase between 3 and 10 degrees Fahrenheit by the middle of the next century. To put that into perspective, the average global temperature has not varied more than 4 degrees Fahrenheit in the 15,000 years since the last glacial period. According to the Union of Concerned Scientists, "If the projections of the greenhouse experts come true, weather can be expected to change in ways *beyond the range of our experience*, bringing extreme abnormalities of heat, cold, drought, and flood."

Reports of the effect of global warming—and the resulting rise in ocean levels due to the melting polar ice caps—are almost too sensational to seem real, but scientists say the scenarios could be very real indeed. If water levels were to rise just 15 feet, much of Florida would end up under water; in Washington, DC, water would inundate National Airport and the Lincoln Memorial and would nearly reach the Capitol steps.

There are other implications of global warming. For example, according to the Worldwatch Institute, rising Earth temperatures could lead to a "wave of extinctions" of the world's species of birds, crops, fish, livestock, plants, and microorganisms.

Some of the greenhouse gases occur naturally. But the bulk of the greenhouse gases are the result of human activity. For example:

• **Carbon dioxide,** responsible for about 50 percent of the greenhouse effect, is created principally by the burning of fossil fuels—wood, coal, oil, and natural gas. Each year, human activity sends some 5.5 billion tons of carbon dioxide into the atmosphere. In the United States, we generate about 6 tons of carbon dioxide per person annually; since 1986, carbon dioxide emissions have grown faster in the United States than in the rest of the world. Another source of carbon dioxide is the burning of trees, particularly the massive burning in tropical rain forests.

• **Chlorofluorocarbons (CFCs),** which account for roughly 15 to 20 percent of the problem, are industrial chemicals manufactured for use as coolants (for refrigerators, air conditioners), cleaning solvents (in computer manufacturing), plastic (as Styrofoam, among many other materials), and foam insulation. CFCs are the principal material responsible for the destruction of the atmospheric ozone layer.

• **Methane,** which contributes about 20 percent of the greenhouse effect, is produced by landfills when organic waste breaks down, as well as in production of livestock (especially plant-eating animals) in farming due to the effects of nitrogen-based fertilizers, and in rice paddies.

• **Nitrous oxide,** accounting for about 10 percent of the greenhouse problem, is formed when certain fertilizers break down in the soil, and by the burning of coal, oil, and other fossil fuels. This substance, which in certain concentrations is known as "laughing gas," is also known to deplete the atmospheric ozone layer.

Other substances, including ground-based ozone smog, water vapor, and airborne particles, contribute to the greenhouse effect.

What You Can Do. There are several things individuals can do to help stem the tide of the greenhouse effect. Among them:

- **Conserve energy.** Because fossil fuels are the primary cause of the greenhouse effect, cutting energy is the primary solution. Insulate your home, buy energy-efficient appliances, buy fuel-efficient automobiles (and keep them tuned), and improve lighting efficiency; lighting and heating alone account for half of all our energy use.
- **Plant a tree.** Every tree takes a certain amount of carbon dioxide out of the air and supplies us with healthy oxygen. Trees also filter out toxins that seep into soil, and prevent them from polluting ground water. If properly planted, trees cool urban neighborhoods, reducing the need for air conditioning. Planting bushes, ground cover, and even window boxes also can help.
- **Buy organic food.** Organic farmers do not use nitrogen-based fertilizers, which have been found to contribute to methane levels in the atmosphere.

3. Ozone Depletion

You've probably become aware in recent years that a sunburn is not just painful—it can lead to skin cancer several years down the road. What would it be like if your risk of skin cancer was increased *every time you went outside*?

That is one of the likely problems resulting from the depletion of the ozone layer of the stratosphere, a thin layer of gas located between 12 and 30 miles above the earth's surface that shields the earth from most of the sun's radiation. Through a seemingly endless list of everyday products and processes, each of us have been unwittingly advancing toward the day when the sun's natural shield will no longer be effective.

The stratospheric ozone *depletion* problem is commonly confused with the ground-level ozone *pollution* problem. In the latter problem, ground-level ozone is created when gases from automobile exhaust and other gases react with sunlight to form photochemical smog. This ozone is the one that causes eye irritation and respiratory problems and that has made life unpleasant for millions of people in and around Los Angeles, among several other cities. Clearly, ground-level ozone is undesirable; we want ozone in the upper atmosphere, not in the air we breathe.

What causes stratospheric ozone depletion? The major known culprits are CFCs—particularly CFC-11 and CFC-12—which have been used in aerosols, in the manufacturing of certain types of plastics, foam insulation, as coolants in refrigerators and air conditioners, and to clean electronic components in the computer industry. These chemicals, whose emissions have grown rapidly since their introduction several decades ago, break apart when interacting with sunlight in the upper atmosphere, releasing chlorine atoms that destroy ozone molecules. Every chlorine atom ultimately destroys several thousand ozone molecules. The rate of ozone-destroying chlorine atoms in the stratosphere is increasing by about 5 percent a year. Even if we were to stop emitting CFCs today, the reactions of existing chemicals would ensure the continued destruction of the ozone layer for at least a century.

We have made some progress toward eliminating CFCs. In 1992, the EPA proposed rules to ban the use of CFCs and other ozone-depleting chemicals in various consumer and industrial products. Most consumer uses of CFCs were banned in 1978. Also in 1992, the EPA issued a rule requiring that CFCs in automobile air conditioners be recycled by service technicians, rather than releasing them into the atmosphere during service and repair. Overall, worldwide production of ozone-depleting chemicals declined about 40 percent between 1986 and 1991.

But such advances have their limits. When automobiles, air conditioners, refrigerators, and other CFC-containing products are junked at the end of their useful lives, their ozone-depleting ingredients often escape into the atmosphere. There are endless millions of these products in existence around the world, each of which must be disposed of properly in order to minimize ozone depletion.

There are indications that the problem is getting worse. A team of scientists discovered in 1991 that the ozone layer over the United States is thinning more rapidly than expected. Because of the lost protection from ultraviolet radiation, Americans face a near-doubling of skin cancer cases and deaths over the next 40 years, according to scientists who analyzed the data.

The problem is not likely to go away soon. CFC concentrations in the atmosphere are expected to grow until the end of the century, reaching a peak 12 to 30 percent above present levels. It will take each of those molecules about 15 years to rise to the level of the stratosphere where the protective ozone layer exists. When broken apart by ultraviolet radiation, each of those molecules will destroy 100,000 molecules of ozone. Each CFC molecules will last about 150 years before it finally breaks down.

Less known but far more destructive to the ozone than CFCs are halons, which are used extensively in fire extinguishers. While not as common as CFCs, halons may have 10 times the ozone depletion potential of the most damaging CFCs. Halon-1211 is used in portable fire extinguishers for industrial locations and in hand-held units for homes and offices. Halon-1301 is mainly used as a fire extinguishing agent, in portable extinguishers in aircraft, and in some sprinkler systems. Most halon emissions do not occur during fire extinguishing, but in testing of the equipment.

What You Can Do. First and foremost is to stop using foam products that contain CFCs. This is easier said than done. Although they are being phased out, CFCs and halons are still involved in the production of countless products. Fortunately, there is a growing list of substitutes for CFCs and halons. DuPont, the largest manufacturer of CFCs, has developed one promising product: a non-ozone-depleting refrigerant called Suva, although it still has some ozone-depleting capability. Several companies have developed flame-retardant chemicals that may be suitable replacements for halons in fire extinguishers.

Among the other things you can do to avoid CFCs:

• **Avoid buying ozone-depleting products.** Despite the elimination of CFCs in many common consumer products, they are still used in others.

• **Use CFC-free insulation.** When adding insulation to your home, don't use rigid foam unless you can verify that it is CFC-free. Good alternatives are fiberglass and cellulose.

• **Use halon-free fire extinguishers.** If you buy a fire extinguisher for your home, purchase a dry-chemical or sodium bicarbonate model; most others contain halons.

4. Air Pollution

Although Americans have expressed concern about air pollution for more than two decades, things aren't necessarily getting better. In 1989, the federal government issued the first ever national survey of toxic chemicals released into the air by industry, which showed that these chemicals are being emitted at rates that threaten public health. According to a 1992 study by the American Lung Association, air pollution may "help to kill tens of thousands of Americans every year."

The problem is by no means limited to big cities like New York and Los Angeles: In the late 1980s, the Environmental Protection Agency cited a nearby coal-fired power plant as the culprit behind the bright white haze obscuring the view of the Grand Canyon. In Florida's Everglades—10,000 square miles of delicate marshland that is home to some of the country's rarest birds and other animals—pollution has made the fish unsafe for eating. The Wilderness Society has predicted that of all the national parks, the Everglades is the closest to extinction. Other parks, including Utah's Canyonlands National Park and Maine's Moosehorn wilderness area, are also plagued by pollution.

Among the biggest threats to human health from the air is ground-level ozone, formed when sunlight interacts with nitrogen oxides and hydrocarbons. Nitrogen oxides come primarily from power plants and automobiles; hydrocarbons emitted by cars, oil-based paints, gasoline (particularly when gas vapors escape while you are filling your tank), dry cleaners, chemical factories, and other sources.

Ozone is used commercially as an industrial bleach. At full strength, it can dissolve concrete and rubber. Even at microscopic levels, ozone irritates breathing passages, causing shortness of breath, coughing, and throat irritation even in healthy people. For children, the elderly, and those with asthma and bronchitis, ozone can be a menace. Ozone is also toxic to crops and trees and may be related to the decline in forests in the eastern United States.

While many of our activities contribute to the problem—with automobile exhaust leading the way—one big villain is American industry, which dumps billions of pounds of toxins into the air each year. The EPA has identified 320 toxic chemicals in the air, only 7 of which are regulated by federal law; 60 of the chemicals have been identified as causing cancer. Most of these are used in various manufacturing processes, including carbon tetrachloride, butadiene, acrylonitrile, and cadmium.

But ordinary citizens, too, increasingly are being conscripted to join the battle for clean air. In Los Angeles, the city with the nation's most polluted air, officials have taken drastic action. The sweeping twenty-year plan adopted in 1989 will not only dramatically curtail automobile use but will also convert most vehicles to the use of nonpolluting fuels by the year 2009. That's just the beginning. The 5,500-page plan also calls for the banning of all aerosol hair sprays and deodorants (not just those containing chlorofluorocarbons), requires companies to install the best antismog equipment available regardless of cost, forces employers to encourage car pooling, and limits outdoor barbecues and gas-powered lawn mowers.

Such seemingly insignificant sources should not be overlooked. Gas-powered lawn mowers, leaf blowers, and chain saws have the pollution-control equipment equivalent to a 1957 Chevrolet—without a muffler. According to the California Air Resources Board, operating a chain saw for two hours emits as many hydrocarbons as a new car driven 3,000 miles. Using a gas-powered mower for an hour generates the pollution equivalent of driving a car for 50 miles.

Los Angeles is only the beginning. We may eventually see similar measures in other polluted and car-choked cities and areas. Automobiles in our major cities account for 40 percent of urban smog. (Another 40 percent comes from dry cleaners, which use solvents that emit fumes, and from bakeries, which release yeast by-products that sunlight changes to ozone.) Cars also emit nitrous oxides, a source of acid rain. While cars have become far less polluting than twenty years ago, there are 55 percent more cars on the road than in 1970, traveling more net miles.

What You Can Do. There are several ways to reduce your contribution to air pollution, most of which have to do with automobile maintenance and use.

- **Don't top off your tank.** When you fill your car with gasoline, fill it only until the pump's automatic shut-off mechanism engages.
- **Use vapor controls when filling your car.** If possible, fill your car at pumps that use devices that capture gas vapors (these are

required at gas stations in California and the District of Columbia).

• **Keep your car in shape.** According to the Environmental Protection Agency, cars get about 40 percent better gas mileage when tuned compared to when untuned. Moreover, a tuned car emits 42 percent fewer hydrocarbons and 47 percent less carbon monoxide.

• **Avoid gas-powered devices.** Opt instead for rechargeable mowers and other appliances.

• **Conserve energy.** Use of energy-efficient appliances and lighting will reduce the running of polluting power plants.

• **Use natural gas.** If your furnace is capable of operating with natural gas, you will be helping to cut back on three major air pollutants: carbon dioxide, carbon monoxide, and sulfur dioxide.

5. Rain Forests and Biodiversity

The world is losing trees—indeed, entire forests—at an astonishing rate. And as the trees disappear, so do thousands of species of other living things. And that may be the least of our problems. The disappearance of trees has been implicated in worsening the greenhouse effect and in air pollution in general.

The most devastating deforestation has taken place in the world's tropical rain forests. Occupying only 6 percent of the earth's surface, these forests are found in warm areas where rainfall is 200 centimeters (about 80 inches) or more each year. This warm, moist environment allows tall, broad-leaved trees to flourish. The trees allow little sunlight onto the forest floor. It is believed that the majority of the earth's species lives in these unique environments, in particular insects and flowering plants.

Tropical forests are being permanently destroyed at an estimated rate of 27 million acres each year, mostly through wholesale burning designed to clear land for agricultural use—including the raising of beef cattle for export to North America. An additional 13 million acres of tropical forest are logged, usually in a fashion guaranteed not to ensure the forests' future. According to a 1991 United Nations report, a total of 42 million acres of rain forest were cleared in 1990, an area about the size of Washington state. "We run the very real risk of squandering a resource base that is fundamental to the future development of the Earth," said Edouard Saouma, director-general of the U.N.'s Food and Agriculture

Organization, in releasing the report. "This situation requires immediate worldwide action."

What does the loss of these trees, insects, and plants have to do with us? Let's start with the trees. In their growth process, trees both store carbon dioxide and convert it into living tissue—wood. In one day a single deciduous tree can absorb about three-quarters of an ounce of carbon dioxide from the atmosphere, roughly 16 pounds a year. When that tree is burned, it not only releases carbon dioxide, it removes one of nature's devices for absorbing it.

As a result, the carbon dioxide released through the burning of tropical forests is thought to contribute at least 20 percent of the global warming problem. (As stated earlier, carbon dioxide is responsible for about 50 percent of the greenhouse effect.) Also released during burning are methane and nitrous oxides, which contribute further to the warming problem.

Of course, not only trees are lost during the burning and logging. Conservative estimates put the number of plant, animal, and insect species native to rain forests at 2 million. Most estimates put the number somewhere between 5 million and 80 million species; one researcher has estimated that there are as many as 30 million insect species in tropical forests alone. Although only a fraction of 1 percent of these species vanishes each year—approximately 4,000 to 6,000 species are lost to deforestation annually—the rate is thousands of times greater than the natural rate of extinction. Indications are that, as deforestation continues, the rate of extinctions could become higher still.

As with the trees, these insects and plants affect our daily lives, and their demise could have immediate economic consequences. In 1970, for example, a leaf fungus blighted U.S. cornfields from the Great Lakes to the Gulf of Mexico, eliminating 15 percent of the entire crop and pushing up corn prices by 20 percent. Losses to farmers exceeded $2 billion. The situation was saved by a blight-resistant germ plasm whose genetic ancestry traces back to variants of corn from one of the plant's native habitats in Mexico.

Moreover, according to Norman Myers, a consultant on the environment and development, half of the purchases in your neighborhood pharmacy, whether drugs or pharmaceuticals, derives from wild organisms. The full commercial value of these wildlife-based products worldwide, says Myers, is $40 billion a year. Remember those poison darts used as weapons by tribesmen in

The Panti Trail

There's nothing like a firsthand glimpse of a rainforest to drive home the nature of the problem. In Central America, the story is familiar: Farmers are slashing and burning rainforests to create farmland, mostly for raising cattle. A small group of activists is trying to stop them, with limited success. Others are trying to work with the farmers.

One of the latter is Rosita Arvigo, a Chicago-born woman who moved to Belize in the mid '70s. Now, with funding from the National Cancer Institute and others, she is collecting information on rainforest plants that may lead to a cure for cancer or AIDS.

Arvigo's Ix Chel farm, down a bumpy dirt road a few miles from the Guatemalan border, features the Panti Medicine Trail, which offers a living display of arboreal and herbal remedies in the forest. But her real purpose there is to run a research facility, not unlike that in *Medicine Man*. In fact, Arvigo has focused her efforts on learning from the region's real medicine men—the few remaining Mayan shamen. Her goal is to preserve their knowledge of the healing properties of native plants. Of the estimated 4,000 plant species in Belize, about 1,500 have so far been found to have medicinal properties.

But her work is a struggle—against time. The forests are disappearing, she says, and while she has managed to arrange with a few farmers to extract some plants before their land is burned, she feels it is a losing battle.

That doesn't hold much hope for the rest of us. Once these species disappear, so, too, does their magic, whatever it may be. And with it goes the possibility that the magic may be there some day when we desperately need it.

Tarzan movies? That "poison," curare, derived from a South American tree bark, is an important anesthetic used during heart, eye, and abdominal surgery. And curare can't be synthesized in the laboratory. Derivatives from the rosy periwinkle offer a 99 percent chance of remission for victims of lymphocytic leukemia, as well as a 58 percent chance of recovery from Hodgkin's disease.

There are thousands of tropical plants that have become valuable for industrial use. First and foremost is rubber. Today, over half the world's commercial rubber is produced in Malaysia and Indonesia. The sap from Amazonian copaiba trees, poured straight into a fuel tank, can power a truck; 20 percent of Brazil's diesel fuel comes from that tree.

And then there is food. Thousands of tropical plants have edible parts, and some are superior to our most common crops. In New Guinea, for example, the winged bean, *Psophocarpus tetragonolobus*, has been called a one-species supermarket: the entire plant—roots, seeds, stems, leaves, and flowers—is edible, and a coffeelike beverage can be made from its juice. It grows like a weed, reaching a height of 15 feet in a few weeks, and has a nutritional value roughly equivalent to soybeans. There are thought to be many such plants that may be developed for commercial use in this and many other countries, creating foods to feed a hungry world.

All told, humans have exploited less than one-tenth of 1 percent of naturally occurring species. And as thousands disappear each year, along with them go their potential.

Agriculture in tropical rain forests is possible, say the experts, if practiced according to the principles of "sustainable use." This refers to any ongoing activity that doesn't permanently degrade the forest, such as the collection of rubber, nuts, or herbs. Environmentalists working with the timber industry have attempted sustainable logging programs, but such plans haven't yet been put into large-scale use.

What You Can Do. It may seem unfeasible to directly affect the burning and harvesting of tropical forests thousands of miles away, but there are actions you can take. Among them:

• **Don't buy wood harvested from endangered tropical forests.** Instead, look for woods that are forested in temperate regions of Europe, North America, and Japan, or are harvested in sustainably managed tropical timber projects. These include ash, beech, birch, cherry, elm, hickory, oak, poplar, and black walnut.

• **Avoid buying beef products from livestock raised in tropical regions.** Indentifying such beef, however, is not always easy. Some rain forest beef is labeled "American beef."

• **Support sustainably managed tropical products.** Examples are nuts and cosmetics made from rain forest herbs. Most such products provide this information on their labels; some are mentioned in the "Food and Groceries" and "Personal Care Products" sections of this book.

6. Garbage

The newspaper and TV news stories a few years back about trash-filled barges and trains with no place to unload their smelly cargoes didn't even begin to describe the extent of the garbage problem we are facing. Some cities are virtually choking on their citizens' refuse, and solutions are few and far between. The simple fact is that Americans generate so much garbage that we are rapidly running out of places to put it.

Although the problem has been around for years, it seems to be getting worse. Between 1960 and 1986, the amount of American garbage discarded annually grew by 80 percent, to between one and two tons of melon rinds, grass clippings, plastic hamburger boxes, and discarded toasters for every living soul in the country. All told, Americans discard 4.3 pounds of solid waste every day, less than a pound of which is recycled, composted, or otherwise recovered. All told, 3.6 pounds per person per day—a total of 195.7 million tons—ends up in landfills. According to the EPA, by the year 2000, our per-capita trash output could reach 4.5 pounds per day, or a total of 222 million tons a year.

The obvious problem is where to put it all. Sanitary landfills—what we used to call "dumps"—are rapidly filling up, sometimes turning away trash generated by those in other regions, as the barge-to-nowhere episode so graphically illustrated. There's good reason for this glut of garbage: 80 percent of all our trash goes into landfills; of the remaining 20 percent, half is incinerated and half is recycled. The U.S. Conference of Mayors predicts that over half of the nation's 9,300 landfills will face closure within ten years.

But where to put the trash is only part of the problem. An equally serious question has to do with what is contained in our trash. Experts believe that somewhere between 5 percent and 15 percent of municipal solid waste contains hazardous substances—garbage that can injure living things and is sometimes even life-threatening. Such hazardous wastes go well beyond the leaky drums of industrial chemicals we've seen on television. They include many of our most common household items.

Another problem with our garbage is that so much of it is made of materials that do not break down easily or quickly. Indeed, we do not even know how long plastic, polystyrene, and similar common substances will last, given that such materials have been invented only in the past few decades. Best guesses for polyethylene, for example, the substance from which disposable diapers are made, is that it will break down in three hundred to five hundred years. But this is entirely speculation; no one really knows.

Even more "natural" products live long lives in landfills. The reason: For things to break down or degrade into benign substances requires sunlight and oxygen, two ingredients in very short supply at the bottom of a trash heap. The tales have been well told of perfectly readable 30-year-old newspapers found in landfills, or long-ago-buried carrots that were still orange. All of which points up the need to find better ways to deal with our unwanted goods.

What exactly is in our trash? According to a 1992 study by the Environmental Protection Agency of municipal solid waste (which includes such things as household trash, grass clippings and other yard waste, used appliances, and fast-food packaging), here is the breakdown, listed by percentage of the total volume:

Paper and paperboard	73.3 million tons	37.5%
Yard trimmings	35.0 million tons	17.9
Plastics	16.2 million tons	8.3
Metals	16.2 million tons	8.3
Glass	13.2 million tons	6.7
Food	13.2 million tons	6.7
Wood	12.3 million tons	6.3
Other	16.3 million tons	8.3
TOTAL	**195.7 million tons**	**100.0**

This list doesn't exactly seem like a toxic chemical dump. Or does it? Here's a sampling of some of what's in our daily trash:

• **Batteries**—used in flashlights, radios, cordless appliances, and dozens of other things—contain cadmium, lead, lithium, manganese dioxide, mercury, nickel, silver, and zinc, all of which are toxic to humans, and all of which can leak out of corroded batteries when they are dumped in landfills.

• **Glass containers, tableware, and cookware** contain lead, which, when it leaks into the water supply, can severely impair children's mental and physical development.

• **Plastics** contain polyvinyl propylene, phenol, ethylene, polystyrene, and benzene, all of which are among the most hazardous air pollutants around.

• **Used disposable diapers** can contain any of over 100 viruses, including live polio and hepatitis from vaccine residues contained in feces.

The Miracle of Rosy Periwinkle

Jay D. Hair, president of the National Wildlife Federation, tells a very personal tale that brings home the value of rainforests:

"About four years ago . . . my older daughter Whitney was very ill with cancer. She literally came within a few days of death. She is here today and she is a beautiful, fourteen-year-old, healthy, completely cured young lady. Why? The drug that saved her life was derived from a plant called the rosy periwinkle. The rosy periwinkle was a plant native to the island country of Madagascar. The irony of this personal story is that ninety percent of the forested area of Madagascar has been destroyed. One hundred percent of all native habitat of the rosy periwinkle is gone forever. And just at a time when we're learning about the marvels of biotechnology. We are losing entire genetic stocks of wild living resources at a time when we're learning about the potential medical marvels of some of these plants, like the one used to cure my daughter. We are destroying them and their potential values forever. This is a tragedy with incredible consequences to the future of global societies."

The tragic fact behind such hazards is that most of what we throw into landfills can be reused, recycled, turned into compost, or otherwise eliminated from the waste stream. Some of it can even be avoided by wiser manufacturing methods and consumer purchasing habits. A sizable portion of our trash contains valuable resources that can be reprocessed and used again.

What You Can Do. There's no doubt that the "three R's"—reducing, reusing, and recycling—are the most important steps in easing the garbage problem. More on that later in this section.

7. Water Pollution

We've come to expect safe drinking water in our homes. Turn on the tap, fill a glass, drink it up, don't give it a second thought. But things just aren't that simple. Despite the years of concern about water pollution, Americans still dump 16 tons of sewage into their waters—every minute of every day. Unfortunately, many of the environmental problems mentioned above—air pollution, acid rain, garbage—end up in our drinking water supply. And even with bottled water, which many people drink to avoid these pollutants, there are environmental problems.

Fresh water is one of the world's most precious resources. Despite the fact that two-thirds of our planet is covered with water, all the fresh water in the world's lakes, creeks, streams, and rivers represents less than .01 percent of the earth's total water. Fortunately, this freshwater supply is constantly replenished through rain and snow. Unfortunately, much of the rain and snow is contaminated by gases and particles that human activity has introduced into the atmosphere.

When we think of water pollutants, we typically picture huge industrial plants spewing millions of gallons of muck into our rivers and streams. While industry does account for a portion of water pollution, treatment of industrial wastes has improved considerably over the past decade. Only about 9 percent of water pollution today comes from direct industrial dumping into rivers, streams, and oceans. Today, concern has been focused on an equally serious problem: groundwater pollution.

If you dig deep enough at almost any spot in the United States, you're likely to hit water. The volume of groundwater within a half mile of the U.S. land surface is said to be four times that of the Great Lakes. Nearly half of all Americans and three-fourths of the urban dwellers rely on groundwater as their primary source of drinking water, using an estimated 100 billion gallons a day.

Underground water picks up whatever it passes through. Rainwater and melted snow—running off parking lots, rooftops, streets, and farms—carry with them deadly substances from worn brake linings, chemical fertilizers, old tires, and a variety of other materials contained in our garbage. During a storm, the pollutants are washed into streams and rivers. Once they get into the water cycle, they never seem to leave.

There are other sources of pollution:

- **Underground gasoline storage tanks**, most of them located at gas stations, leak petroleum products into the soil. The federal government has estimated that there may be as many as 100,000 leaky gasoline storage tanks in the United States.
- **Agricultural irrigation**, which accounts for about two-thirds of all water use, can leach salts and pesticides from the soil and into the water. The Environmental Protection Agency has estimated that each year between 3.5 million and 21 million pounds of pesticides reach groundwater or surface water before degrading.

• **De-icing salts**, used throughout the country's frost belt, have caused contamination in eleven northern states as they dissolve and percolate into the soil.

One possible solution might be to drink bottled water, assuming that everyone could afford that luxury. But most bottled water, despite advertising and marketing campaigns that might suggest otherwise, isn't necessarily more healthy than most cities' tap water; it just tastes better. The regulations currently enforced by the Food and Drug Administration set standards for bottled water that are very similar to the EPA's standards for tap water. And the tap water standards are not known for being strict. The rules set "tolerance" levels for some very potent chemicals, but they don't specify any limits for a wide variety of synthetic compounds called "organics"—an alphabet soup of carcinogens and mutagens such as polyvinyl chloride, trihalomethanes, and polybrominated biphenyls (PBBs). Around 700 organics have been measured in drinking water so far; many of them have never been tested for toxicity.

In addition, bottled water, however tasty, encourages a wasteful and polluting industry. For one thing, the energy involved in moving water around the world in bottles is appalling. In addition to transportation fuel, there is the cost of packaging. It takes the energy equivalent of a ton of coal to produce a ton of glass; plastic bottles, of course, contain other notorious pollutants; about a third of bottled water is sold in nonbiodegradable and nonrecyclable plastic containers. Moreover, the plastic polyethylene containers are known to leach their chemicals into the water, which mixed with other impurities, can be toxic. Another solution—investing in a household water-filtering device that is connected to your tap—may be effective (although some filters create health hazards themselves; see "Fresh Drinking Water" for details), but it can be costly. The best solution is to work toward improving the existing water supply in your community.

Beyond the issue of water pollution is the question of water supply: Is there enough water available for everyone? The good news is that there is plenty of fresh water on the earth—about 9,000 cubic kilometers—which, according to one estimate, is enough to sustain 20 billion people, more than four times the earth's population. The bad news is that the water and its people are unevenly distributed. So some areas (like Iceland) have far more water than they can possibly use; many other areas have far less than they need.

In the United States, there are fewer extremes. While some areas face water deficits, there are relatively few areas with excess water supply. An increasing number of communities are turning to water conservation as a means of making scarce water resources available to everyone or of ensuring that adequate water supplies remain reliable.

Besides ensuring an adequate supply, there is at least one additional benefit to conserving water. Water conservation helps protect fish and wildlife habitats and wetlands. These environments may be irreparably damaged if there is not enough water to sustain them, with serious ecological consequences and, in some communities, economic impacts as well.

By diverting less water, we leave more water for other uses, from boating and fishing to power generation. Maintaining adequate flow also helps ensure water quality: Water is more susceptible to contamination when there is a low flow. And by using less water we also discharge less used water, thereby further reducing the potential for pollution. Finally, using less water minimizes the need to build new dams and reservoirs, costly projects that can further destroy wildlife.

What You Can Do. There are two areas of concern: conserving water and keeping it clean.

• **Don't waste water**. Much of the water we use at home is wasted. From installing water-saving devices to making changes in your daily habits, there are many ways to conserve water without compromising convenience.

• **Don't add pollutants at home**. Some toxic products—such as bleaches and paints—should not be dumped directly down the drain. There are other environmentally safe methods of disposal.

• **Buy products that don't pollute the water**. Some soaps and detergents, for example, contain phosphates and other substances that can harm the water supply. Fortunately, there are safe alternatives to most of these products.

• **Report water polluters**. If you believe a factory or other business in your area is dumping harmful substances into local lakes or streams, report them to local or federal officials; additional information is contained in Part III of this book.

THE PROBLEM OF PACKAGING

We have become the most overpackaged society in history.

We love packaging. And the products in our supermarkets, drugstores, and hardware stores reflect that love affair: nearly everything, it seems, is wrapped in something. Even produce—tomatoes and corn on the cob, for example—sit neatly on a plastic foam tray, encased in clear plastic wrap. Some products have layers upon layers of wrapping, for no apparent practical reason. According to a 1992 EPA study, packaging comprises just under 30 percent of all solid waste. And the overwhelming bulk ends up in landfills or incinerators. Excluding corrugated cardboard boxes, which are widely recycled by businesses, less than 10 percent of all other paper, glass, steel, aluminum, and plastic packaging trash is recycled or otherwise recovered.

For the nation, the statistics about our packaging trash mount up into some pretty impressive numbers. For example, according to the Environmental Defense Fund:

- We throw away enough glass containers to fill the 1,350-foot twin towers of New York's World Trade Center *every two weeks*.
- We toss out enough aluminum to rebuild our entire commercial air fleet *every three months*.
- We go through 2.5 million plastic bottles *every hour*.
- We discard enough iron and steel to *continuously* supply all the nation's auto makers.

Guess who's paying for all this waste? We are, in a number of ways. Packaging not only adds to the price of these products, it also costs us money to dispose of it or to repair the harmful effects many of the packaging materials have on the environment.

Make no mistake: Packaging is important. It protects products from damage, ensures that they remain sanitary, prevents tampering, provides space for product information, and offers convenience. But the manufacture, use, and disposal of packaging materials contribute to many environmental problems, from litter to acid rain.

The New Generation of Packaging. The problem isn't just the amount of packaging, it's also the types of materials being used. A growing number of products are being wrapped in "composites"—packages containing several layers of materials and adhesives, such as aseptic juice boxes, which contain layers of polyethylene, paperboard, and aluminum. Squeezable ketchup bottles, made of up to seven layers of plastics and adhesives, are another example. When these bottles are empty, you cannot separate their various materials for recycling, so they are guaranteed to end up in landfills, where they will last for centuries, or incinerators, where their ash may contain toxic materials that can get into the air, soil, and water.

Even when packaging consists of only one type of material, it is often an unrecyclable one. The vast majority of Americans have no means to recycle most types of plastic or polystyrene, or even the kind of coated paperboard used on cereal and cracker boxes and many other packages. Many manufacturers, trying to lure environmentally conscious consumers, claim their packages to be "recyclable." As stated earlier, while technically true, it is not true for most consumers.

The environmental problems of packaging aren't limited to what happens to it when you throw it away. The manufacturing process is also key. The amount of air and water pollution created during the package's manufacture, the amount of energy required to transport it to market, the need to keep it refrigerated or frozen (requiring more ozone-depleting coolant)—all are factors in determining which material or combination of materials comprise the best types of packaging.

And then there are the inks used in printing the packages' colorful labels. Some of the inks contain heavy metals—toxic elements such as lead, cadmium, and chromium. Although studies are inconclusive, some researchers believe that heavy metals may leach into groundwater when packages are discarded in landfills, or that they may contaminate the ash that spews from incinerators when the packages are burned. Heavy metals are just the beginning. Dyes, solders, and other additives are under scrutiny for their possible environmental impact.

Definitive information about all these issues is elusive. While each industry—glass, paper, aluminum, plastic, polystyrene, and juice boxes—boasts impressive studies supporting the notion that theirs is the material of environmental choice, there is little consensus, even among independent scientists.

The Best and Worst Packaging

In reviewing this list, keep in mind that a key component is the type of material that can be readily recycled in your community. If you live in a community in which plastic soda bottles or polystyrene can be recycled, for example, your list of "best" and "worst" will differ from someone who cannot recycle these things.

1. Best. What's the best type of packaging? Easy: no packaging at all. You'd be surprised how many products don't need any packaging. Is it really necessary to shrink-wrap produce on a cardboard or Styrofoam tray? Probably not. (In fact, you probably don't even need plastic produce bags for fruits and vegetables with their own peels or rinds. You throw five cans of soup into your grocery cart and shopping bag, why not five oranges?) Many nonfood items also can be sold perfectly well without packaging. Where sanitation and security are not a concern—as with hardware and variety store items—packaging may not be needed.

2. Very Good. When some packaging is required, the rule is: Use the least amount of packaging possible, using the highest content of recycled and recyclable material. So, a glass bottle with an aluminum or steel top—all of which are easily recyclable—is perfectly adequate. An aluminum, steel, or bimetal can is fine, too. (A bimetal can combines aluminum, tin, and steel; they are used to package many foods.) Seeking minimal packaging also leads one to buy the largest-size package possible. Several smaller packages inevitably mean a greater amount of packaging materials for the same amount of product. Another way to minimize packaging is to buy one of the growing number of concentrated products, especially detergents and other cleaners; simply put, they combine more product in less space. Some brands combine detergent and bleach, thereby halving the number of packages needed to do laundry. Both are good strategies for cutting packaging.

What about cardboard? Many products, including most cold breakfast cereals, are packaged in materials made from recycled paper or cardboard collected from large, industrial users of these materials. That's the good part. But many recyclers do not accept these for recycling, so they will end up in incinerators or landfills.

Also ranking high are endlessly refillable packages—namely

returnable glass beverage bottles. Egg cartons, assuming they are returned and reused, also rate high.

3. Good. Reusable, refillable, and recyclable packages are a step in the right direction. Margarine tubs and other containers that can be used for food storage are good examples, although there are limits to the number of these most households can put to use. Another innovation are products in which you purchase the big package once, then buy smaller refills from there on. Concentrated Downy Refill, for example, which comes in a nonrecycled and nonrecyclable milk cartonlike container, can refill a hard plastic jug of the original product; you simply pour in the concentrate and add water, like making frozen orange juice. Pillsbury sells a microwave cake mix that includes a reusable plastic pan. After buying the pan, you need only buy refills from there on. Eventually, of course, the cake pan—and the empty fabric softener jug—will need to be discarded. Neither type of container is easily recyclable.

Of course, anything that you can recycle is desirable. But as stated previously, those materials vary from community to community, even neighborhood to neighborhood. Better still are recyclable packages made from recycled material.

4. Fair. In the middle ground are packages made from recyclable, but not recycled, material. (Again, this has a lot to do with what's recyclable in your community.) Less desirable are packages made from recycled materials that cannot themselves be recycled; they are at the end of their useful life and destined for a landfill or incinerator. Still okay but barely acceptable are containers made from at least some recycled material. Some plastic jugs, for example, contain up to 25 percent recycled plastic. You can find polystyrene egg cartons also containing one-fourth recycled polystyrene.

5. Bad. Products packaged in several layers of wrapping are usually not desirable, and their desirability drops in direct proportion to the amount of unrecycled and unrecyclable material used. Supermarket shelves are being increasingly filled with such products, mostly convenience and microwave foods. But not exclusively: we found overpackaged frozen foods, snacks, beverages, cleaners, baby products, and personal care products.

Equally undesirable is single-layer packaging made from

Packaging, Best to Worst

Best	• No packaging at all
Very Good	• Minimal, recycled, and recyclable packages • Concentrated products • Endlessly refillable packages • Reusable, refillable, and recyclable packages
Good	• Packages made of recyclable, but not recycled, material • Packages made of recycled, but unrecyclable, material
Fair	• Packages made from partial recycled content • Products packaged in multiple layers of recyclable packaging
Bad	• Products packaged in multiple layers of unrecyclable packaging • Single-serving containers • Aseptic packages • Aerosol cans
Worst	• Packages made of composite materials

unrecycled and unrecyclable materials. Again, specific products in this category have a lot to do with what you can recycle in your community. For some people, plastic soda bottles and milk jugs will fall into this undesirable category because they cannot recycle these things. For those who can recycle these plastics, such beverage containers would probably fall into the "Fair" category above.

Some of these "Bad" products look innocent enough. A white cardboard box surrounding a box of donuts, for example, may not appear villainous, but that packaging may take up as much space in a landfill as a polystyrene meat tray, may be just as polluting to manufacture, and may take nearly as many centuries to degrade.

6. Worst. At the bottom of the heap—literally—are single-serving containers, especially when they are made from multiple layers of materials, also known as "composites." Single servings, no matter what they're made of, create a great deal of packaging waste for the amount of product delivered.

It's not that Green Consumers must abandon convenience in their daily lives. But there are greener alternatives: resealable plastic containers can hold small servings of just about anything, including beverages. Moreover, if necessary, you can place these containers in a microwave oven for cooking. (**Beware**: *Remove the tight-fitting plastic top and cover with wax paper before cooking; otherwise, heat will not be able to escape and the container will melt.*) An investment in a set of such containers will more than pay for themselves through the cost savings you'll get from buying large-size packages, instead of single servings.

Multiple Materials. Composites are particularly troublesome. Squeezable ketchup and mustard containers, for example, made from several layers of plastics and adhesives, are simply not recyclable, even in communities that have the ability to recycle several kinds of plastic. They are guaranteed to end up in landfills.

Another environmentally undesirable container is the aseptic package, also known as the juice box—those small boxes of juice and flavored drinks, accompanied by their own plastic straw. Juice boxes are made from three layers of materials—70 percent coated paperboard, 23 percent polyethylene plastic, and 7 percent aluminum foil. While they're lightweight and less voluminous, taking up minimal space in a landfill, they are not readily recyclable. (The aseptic industry has managed to turn some juice boxes into a kind of plastic lumber, but this is a demonstration project only and currently has no economic feasibility for wide-scale use.) Again, there are alternatives, including small resealable plastic juice containers that can be washed and reused. By buying larger sizes with less packaging, you'll get far more beverage per dollar—as well as doing something good for the environment.

Aerosols. Spray cans are another undesirable packaging type. Many people believe that because chlorofluorocarbons were banned in 1978, aerosol cans are now environmentally safe products. But most aerosols aren't environmentally safe. For starters, many substitute propellants—which force the hair spray or deodorant out of the can—are hydrocarbons, such as propane or butane. Just like the hydrocarbons that come out of your car's tailpipe, these mix with sunlight to form photochemical smog. For that reason, Los Angeles will soon be regulating the sale of things like aerosol hair spray and deodorants. In addition, some common household aerosols contain toxic solvents that end up in the air and in our lungs.

And aerosol packaging itself is unrecyclable in the overwhelming majority of communities because it is difficult to separate the various components of the can.

There have been some improvements in aerosol technology in recent years. A few manufacturers have introduced products that use pressured air as a propellent. With these products, you must first pump the top or bottom by hand a few times to build up pressure. Then you spray in the normal fashion. At least one company also sells a refillable version, which allows you to use the aerosol device over and over; you can refill the container with anything from perfume to paint.

Making the "right" packaging choices can be extremely difficult. If you have kids, for example, it may be difficult to give up juice boxes. But do what you can. By starting somewhere, you will be making a small but meaningful contribution to reducing our overflowing landfills and helping to make the most of our resources.

The Perils of Plastic

There is no question that plastic is the packaging material of choice for many product manufacturers. Since the 1940s, manufacturers have come to depend on a seemingly endless variety of plastic packaging materials: bottles, bags, jars, jugs, tubes, tubs, foams, films, pellets, wraps, and so on. Today, the volume of plastics used in the United States exceeds that of steel. The plastics industry predicts that consumption of plastic resins will grow an estimated 50 percent during the 1990s. The single largest use of plastics today is packaging, constituting a fourth of the 12 billion or so pounds of plastics produced each year in this country.

The debate over plastic is an intense one. Both environmentalists and the plastics industry generate endless reams of paper—recycled, we hope—on the subject, and each side has well-crafted arguments. Indeed, if you were to listen only to the plastics industry, you would be led to believe that its miracle material has helped to save the environment. One industry group, for example, provides reams of data "proving" that, compared to glass and aluminum, PET, the plastic used for beverage bottles, is more energy efficient, creates less pollution, and is fully recyclable. Environmentalists, of course, have a much different point of view.

But the use of plastics has a two-pronged effect on the environment. Many of the chemicals used in the production and processing of plastics are highly toxic. Most of the plants that produce these chemicals also produce hazardous wastes that pollute air and water. In 1986, when the EPA ranked the 20 chemicals whose production generates the most hazardous waste, five of the top six were chemicals commonly used by the plastics industry: propylene, phenol, ethylene, polystyrene, and benzene. Most of the ten biggest-polluting companies in the United States (based on Toxic Release Inventory data reported to the federal government) are primarily in the plastics and chemical business.

The effects of plastics can be found beyond our air and water. They also affect many living creatures. Floating plastic debris can be found in the world's oceans. One marine study group concluded that the garbage floating in the world's seas now outweighs the fish harvest by a factor of three to one. Plastic six-pack yokes, Styrofoam pellets, plastic bags, and tampon applicators are among the more common materials bobbing in the waves. The thousands of fish, birds, and marine mammal species that feed off the water surface cannot distinguish these things from food. Indeed, much of it resembles their very diet: floating plastic bags are mistaken for jellyfish; bits of polystyrene foam resemble fish eggs; cigarette butts look like small brown crustaceans. The results are tragic, with wildlife choking, starving, or developing lethal infections from this sea of trash. A report from the federal Office of Technology Assessment concluded that plastic pollution is a greater threat to marine mammals and birds than are pesticides, oil spills, or contaminated runoff from land.

Burn, Bury, or Recycle?

Getting rid of plastics isn't easy. There are basically three choices:

- **Burning** plastics in incinerators is a risky business. Most plastics release a host of toxic materials when burned, including such heavy metals as cadmium, lead, and nickel. These substances are emitted into the air or are contained in the ashes left behind—which still must be disposed of somehow.

Which Plastic Is It?

How can you tell the kind of plastic used in a package? It can be very hard to do. Experts say it's nearly impossible to tell without conducting a a chemical analysis.

Some help comes from a coding system created by the plastics industry, designed to help recyclers identify which packages could be recycled. If you can recycle plastic in your community, this system will help identify packages that are recyclable. (If a container has no code, you can safely assume that it is made of several types of plastic, and is therefore not recyclable.) Please understand: *This system does not ensure that your plastic purchase will be recycled. It is simply an identification system.*

The system includes seven codes, one of which is usually stamped on the bottom of a package. Please note that of these seven plastics, only two of them—PET and HDPE—are currently being recycled in any quantities. Here are the codes:

1 PETE

1. Polyethylene terephthalate (PET or PETE)—used mostly in soda bottles, also found in meat containers, cosmetic packages, and boil-in bags.

2 HDPE

2. High-density polyethylene (HDPE)—commonly used in such consumer packaging as milk and water jugs, and shampoo and detergent bottles.

3 V

3. Polyvinyl chloride (PVC or V)—used to package floor polishes, shampoos, edible oils, mouthwashes, and liquor.

4 LDPE

4. Low-density polyethylene (LDPE)—used when squeezability is desired, such as in toiletries and cosmetics.

5 PP

5. Polypropylene (PP)—used to package foods that must be packaged while hot, such as pancake syrup.

6 PS

6. Polystyrene (PS)—used in egg cartons and to package tablets, salves, ointments, and other products not sensitive to oxygen and moisture.

7 Other

7. Other plastics—includes a wide range of substances, including mixed plastics, which cannot be recycled.

• **Burying** plastics, of course, has its own problems, not the least of which is the shrinking amount of landfill available to hold a growing amount of plastic trash. Despite some early claims of "degradable" plastics designed to disintegrate or decompose over time, such technology doesn't yet exist. As we said earlier, in landfills, where most plastics end up, practically nothing degrades.

Even when plastics do degrade, questions remain. For instance, what materials do plastics degrade into? The liquid that filters through degradable waste, called "leachate," picks up toxic substances, such as heavy metals used as pigments and stabilizers in plastics, which can subsequently contaminate groundwater.

• **Recycling.** That leaves the third option, recycling, one that is being promoted aggressively by the plastics industry. As the Plastics Recycling Foundation, an industry group, put it: "We fervently believe that plastics hold the potential for becoming the most recyclable packaging material."

But most recycling plastic is different from recycling glass, aluminum, cardboard, or paper. While the recycling process for these materials is continuous—you can recycle an aluminum can into another aluminum can or a glass bottle into another bottle—most plastic recycling is a one-time thing. Researchers have developed processes to shred plastics—primarily from soda bottles—into flakes, which are then washed and separated from residual materials. The plastics recycling industry has found many ingenious uses for this shredded plastic: furniture cushions, bathtubs, flowerpots, pipes, toys, trash cans, shipping pallets, parking lot car stops, boat docks, and insulation, among a growing number of other products. But these are mostly one-time uses. All of these products have limited lifetimes, and they, too, will need to be disposed of eventually. It is unclear whether there will be any technology to take these parking lot car stops, flowerpots, and other items to reuse their plastic once again.

There are two big obstacles to plastic recycling. One is that federal law has restricted the use of recycled plastic for food packaging. Only a few products—namely, soda bottles—contain recycled plastic, at rates typically no more than 25 percent.

A bigger problem has to do with the economics of plastics recycling. The glut of oil on the world market has depressed the prices of the petroleum-based chemicals from which plastics are made. That makes it cheaper to manufacture plastics from virgin chemicals than from recycled materials. Plastics recycling is an expensive proposition, partially because of the material's huge bulk and light weight. A huge truckful of plastic bottles often doesn't bring a high enough price to warrant shipping it hundreds of miles away to a recycling plant.

So, plastics recycling in most cases is not yet a cost-effective process. And the outlook for improved economics in the near future isn't optimistic. While recycling of some plastics will continue to grow, the overwhelming bulk of plastic waste will continue to end up in landfills and incinerators.

Reduce, Reuse, and Recycle

What, then, is the solution? As with so many other environmental problems, the solution requires a varied, integrated approach. There are three key parts:

- **Avoid plastic.** The perfect solution is to seek alternatives to plastic packaging. While you can't avoid it entirely, there are many products packaged in materials that are more easily recyclable in most communities.
- **Reuse plastic products.** Whenever possible, reuse plastic containers—for storing food, hardware, or other small items. Also, look for products in containers that can be refilled without your buying an additional plastic container.
- **Recycle plastic products.** If you must buy plastic, seek out those that can be recycled. Keep these plastics separate from other trash, then bring them to a recycling center that accepts plastics.

As you can see, there is no simple answer to the plastics problem. For now, plastic packaging is a reality in our lives. But the less plastic packaging—or any packaging, for that matter—the better. The best solutions for Green Consumers are those that use—and reuse—plastics in a responsible way.

THE RECYCLING SOLUTION

Let's start with the basics: Recycling makes sense. It is both ecological and (usually) economical. For Green Consumers, no activity has a more direct payoff to the environment. Problems from water and air pollution to landfill overflow and the greenhouse effect—in other words, just about every environmental problem faced in the 1990s—can be eased through recycling.

Recycling is not a new idea. Far from it. Half a century ago, during World War II, recycling was a way of life for many Americans. Everything from tin cans to rubber to cooking grease was recycled to help the war effort. Everyone did his or her part to preserve the country's scarce or strategic resources. When the war ended and manufacturing boomed, Americans developed a throwaway mentality. Besides, America is such a large country that no one figured we would run out of space to bury our trash.

Now, reality has set in, and recycling has come full cycle. America's scarce and strategic resources may have changed, but the solution to preserving them has remained the same: recycling.

Consider just a handful of the benefits to the environment recycling can bring:

- Using recycled instead of virgin paper for one print run of the Sunday edition of *The New York Times* would save 75,000 trees.
- The amount of copper thrown away each year—primarily from electronics—in a city the size of San Francisco is more than the amount produced annually by a medium-size copper mine.
- Recycling creates six times as many jobs as does landfilling and incineration.
- In 1990 alone, recycling of aluminum cans saved more than 12 billion kilowatt-hours of electricity, roughly enough to supply the residential electric needs of New York City for six months.

These impressive statistics notwithstanding, Americans currently recycle only about 14 percent of their household trash. In the process, they throw away incalculable amounts of raw materials—and the energy it takes to convert these materials into finished goods. Part of the problem may be that many people don't understand what can and what cannot be recycled. According to a poll by the Gallup Organization, Inc., 54 percent of Americans said that if they knew a certain food or beverage container was not recyclable, they would switch to a container that was.

There are at present some 8,000 community recycling programs in the United States. These range from voluntary drop-off programs, in which individuals bring recyclable materials to central locations, to mandatory curbside pickup programs, in which citizens are required by law to separate their garbage and prepare it in a prescribed manner for pickup by the local trash company. In addition, a growing number of business recycling programs are emerging, including some mandated by state or local law, in which companies save, reuse, and recycle the massive amounts of paper, cardboard, metals, and other materials used in our information and manufacturing society.

Recycling is really only part of what has become known as "integrated waste management"—a bureaucratic term that refers to a four-part strategy for effectively dealing with landfill and pollution problems: *reducing* the amount of trash generated; *incinerating* refuse, preferably turning the heat into energy; *recycling* as much as possible; and *landfilling* whatever is left. The federal government has been promoting these integrated programs, which can be made flexible enough to meet each community's needs.

What's Recyclable?

What exactly is recyclable? Here, in alphabetical order, are the commonly recycled products and materials.

Aluminum. Aluminum is one of the most expensive and polluting metals to produce. For one thing, the metal is extracted from bauxite ore, which is mined on the surface, much of it in tropical rain forest areas. Fortunately, recycling aluminum cans is one of the great environmental success stories of recent years. Since the early 1970s, can recycling has grown steadily; Americans now recycle about 55 percent of all aluminum cans—some 42 billion recycled cans containing about 1.5 billion pounds of aluminum. According to the Aluminum Association, that represents more than 5 million tons of aluminum that has not been buried in our landfills since 1972.

The aluminum recycling process is relatively swift: Used aluminum cans are melted down and returned to store shelves in the form of new beverage containers in as few as six weeks. One reason for such speed is the ease of recycling an all-aluminum can: Because most aluminum cans have no labels, caps, or tops, they needn't be separated from other "foreign" materials before being reused.

Nearly all recycling facilities accept all-aluminum cans. (To tell the difference between an aluminum can and a steel one, apply a magnet. If the magnet sticks, the can is steel.) Some centers accept bimetal cans—those with aluminum top and steel sides. In addition to cans, you may include aluminum foil, pie pans, and other aluminum scrap. You needn't prepare aluminum for recycling except to separate it from other trash. For your convenience, however, it might help to crush cans before discarding. Of course, in many areas, you can return the cans to stores for a refund.

Batteries. The number of dry-cell batteries we use is growing by nearly 10 percent a year. We now throw away about 2.5 billion batteries—from the cylindrical flashlight variety to the tiny button-size type used in cameras, watches, calculators, and hearing aids.

These innocent-looking batteries are rarely considered environmental threats, but they contain some very toxic chemicals. When burned in incinerators, the heavy metals in batteries—including cadmium, lead, lithium, manganese dioxide, mercury, nickel, silver, and zinc—pollute the air or become toxic components of discarded incinerator ash. When tossed into landfills, the metals leach out of corroded batteries and seep into the groundwater. These metals are so dangerous that the Occupational Safety and Health Administration has established workplace exposure limits for all eight metals mentioned above.

Battery recycling is done routinely in Europe and Japan, but it has only begun to catch on in the United States, which has only a handful of battery recycling programs.

Another solution comes in a new breed of mercury-free disposable batteries. While they're still thrown away, at least they lack one of the most toxic ingredients. **Eveready Battery Company** is one of several companies that have introduced an alkaline battery that is 99.975 percent mercury-free.

Perhaps the best solution are rechargeable batteries. Although these batteries cost more and require the one-time added expense

No Accounting for Taste

A food scientist at Oregon State University seems to have come up with the ultimate in "green" packaging: edible wrappers. Antonio Torres has developed an edible coating made from natural ingredients. The tasteless and colorless films are made from corn protein and cellulose; both are approved by the federal government for food packaging. A third coating is derived from the casings of shellfish. When sprayed onto foods in layers as thin as a few thousandths of an inch thick, the coatings prevent spoilage for up to a year. Several food companies are interested. One of the first applications: creating a layer between ice cream and cones to prevent the cone from becoming soggy. That, undoubtedly, will lead to pre-scooped ice cream cones in your supermarket's frozen food section.

of a recharger, most such batteries can be recharged up to one thousand times. After that, the nickel and cadmium can still be reclaimed and recycled. There are even solar-powered battery charging systems, which use sunlight to recharge batteries. It takes about six hours to recharge a set of batteries. See "Home Energy and Furnishings" for more on rechargeables.

Corrugated cardboard. Corrugated cardboard—used primarily in cardboard boxes— like paper, is fully recyclable. Cardboard, in particular, is in short supply, so the demand for used cardboard is high. About half of all corrugated cardboard used in the United States is recovered each year, and the percentage is expected to rise in coming years.

To recycle cardboard, you need only to flatten boxes and tie them together for drop-off or pickup.

Clothing. Many people overlook the obvious: the clothes on their back. While there are no reliable figures on the amount of reusable worn clothing tossed out each year, common sense would dictate that any amount is a waste, particularly in light of the millions of Americans who cannot afford to shop for clothes. Humanitarianism aside, these thrown-away clothes represent a tremendous amount of discarded energy and resources.

Almost every community has a local service or religious organization that will pick up used clothing—as well as most other household goods—and distribute them to those in need. Some, but not all, of these organizations repair broken or torn items before distribution.

Glass. Like aluminum, glass is 100-percent recyclable. Glass makes up about 10 percent of household garbage and is one of the easiest of materials to recycle. The broken glass (known as "cullet") is added to new molten glass in a furnace, producing new glass. Through the use of cullet, which melts at a lower temperature than the raw materials used to make glass, new glass can be made with considerably less energy and resources than would otherwise be needed. Glass can be reused an infinite number of times. Of course, some bottles needn't be melted down at all. Like milk bottles of years gone by, they are merely washed, sterilized, and refilled.

There are three basic types of glass: clear, green, and brown. Not all recycling centers accept all three types, but some do. In any case, you should separate the different-color glass into different containers (or at least separate out the clear glass, if that's all your local recycling center will accept). Broken glass is perfectly acceptable, and while you needn't remove paper labels, you may need to remove aluminum neck rings and caps; your recycling center will provide details. Some types of glass aren't recyclable, however. Most recycling centers, for example, will not accept light bulbs, ceramic glass, dishes, or plate glass, because these items consist of different materials than do bottles and jars.

Paper. The United States leads the world in paper consumption—and trails far behind in paper recycling. Americans use some 67 million tons of paper annually, recycling about 30 percent of it. This statistic is particularly distressing in light of the fact that the production of a ton of paper from discarded waste paper requires 64 percent less energy, needs 58 percent less water, results in 74 percent less air pollution and 35 percent less water pollution, saves 17 pulp trees, reduces solid waste going into landfills, and creates five times more jobs compared to producing a ton of paper from virgin wood pulp. According to the Institute of Scrap Recycling Industries, more than 400 million trees are saved each year due to current recycling efforts. Paper makes up nearly 40 percent of municipal solid waste by weight.

The lack of paper recycling in America has less to do with people's unwillingness to recycle than with the lack of recycling mills in operation. While the eight mills in the United States capable of recycling newspapers run at full capacity and sell all they produce, the price tag for building a new recycling mill begins at a whopping $400 million. Considering the fluctuating price of paper—the demand for newsprint, for example, is directly related to the demand for advertising linage, which is linked to the economy as a whole—investors are reluctant to take such a risk.

But the demand for recycled paper exists, and all forecasts indicate that as the demand continues to grow, new manufacturing plants will open.

Not all paper is recyclable. Newsprint is among the most desirable papers, but recyclers are picky even about that. Coated papers—such as those used for Sunday magazines and newspaper advertising supplements—can make sheets of rolled paper stick together. Yellow paper, including legal pads, is also undesirable. Brown bags and junk mail also gum up the works. The widespread assumption that brown paper is always recyclable simply isn't true. It must first be separated from other types of paper.

A few state-of-the art plants, however, can recycle many of these types of paper, as well as envelopes with plastic windows, greeting cards on glossy stock, and other things. The Fort Howard Corporation in Wisconsin, for example, uses old magazines, phone books, Post-its, and many other types of supposedly unrecyclable paper to produce its two lines of paper towels, toilet paper, and napkins.

Most office papers may be recycled too. Office workers produce an average of a pound of waste paper a day, according to the Office Paper Recycling Service, a New York consulting group. (Ironically, much of the paper is generated by computers and other equipment that was supposed to represent the paperless office of the future.) Unfortunately, the common yellow legal pad is not accepted by most paper recyclers because the dyes make it difficult to produce the bright white paper most consumers want; other paper colors are equally undesirable.

With the demand for recycled office paper being much greater than the demand for newsprint—and, therefore, the price for recycled office paper being considerably higher than for newsprint—many large companies have recognized that recycling office paper is as profitable as it is environmentally sound. According to published reports, the American Telephone & Telegraph Company saved $1 million in disposal costs in 1988 and made a $365,000 profit by recycling its office waste paper. Other companies report similarly impressive figures. The World Trade Center in New York City set up a program for its 50,000 workers. It is expected to save 262,500 gallons of water, 153,750 kilowatt-hours of electricity, 371 cubic feet of landfill space, and 637 trees—every day.

Oil. While an increasing number of people are changing their own motor oil, many of them—as well as a number of professional service stations—do not dispose of the used oil properly. And if you think that the few quarts of oil draining from your car's crankcase won't hurt anyone, think again: As little as a quart of oil, when completely dissolved or dispersed in water, can contaminate up to 2 million gallons of drinking water. A single gallon can form an oil slick of nearly eight acres. Used motor oil poured down the drain or into the nearest storm sewer often goes directly to the nearest creek, river, or lake, killing aquatic life and polluting drinking water.

Many communities have begun curbside collection programs to recycle used oil. Used oil is a valuable, renewable resource, although it must be handled carefully. Through refining, about two and a half quarts of new motor oil can be extracted from one gallon of used oil. When recycled, used oil is reprocessed with water, then used for fuel. About a fourth is re-refined and turned into base oil stock. With additives, it is used as lubricating oil and put to other uses. Some of it is used for road oil, dust control, wood preservatives, and "fire log" ingredients. The production of re-refined oil uses just one-fourth the energy of refining from crude oil. According to the U.S. Department of Energy, Americans use about 1.2 billion gallons of oil annually—about 78,000 barrels a day. About 60 percent of it (some 700 million gallons) is used motor oils. The remaining 500 million gallons are industrial oils.

Plastics. As stated earlier, plastics recycling is a growing—and controversial—process. Researchers have developed processes to shred plastics, primarily from soda bottles, into flakes, which are then washed and separated from residual materials and made into a wide range of materials. But unlike paper, glass, and aluminum, at present, plastics can be recycled a limited number of times. Moreover, only a relatively small number of recycling centers and even fewer curbside pickup programs currently accept plastics. If you can recycle plastic, it is best to rinse it, then crush it to save valuable space in the vehicles that haul the plastic to be recycled.

Steel Cans. All steel food and beverage cans are 100 percent recyclable. Yet, each year, about 30 billion steel cans with a thin tin coating are dumped into America's landfills. The technology to reclaim and recycle these two materials has been around for more than sixty years, and the capacity to recycle far exceeds the availability of recyclable cans. You can easily determine which cans contain steel with a simple magnet: if the magnet sticks, it's steel; if not, it's probably aluminum.

Tires. We throw away 220 million tires a year, millions of which end up in huge unsightly—and fire-prone—stockpiles around the country; currently, well over a billion discarded tires sit in such mounds. But tires can be recycled in a number of ways: almost 40 million tires are retreaded annually; another 10 million are shredded, then used for sheet rubber, asphalt-rubber for roadbeds, and other products; some shredded tires are used as fuel to generate electricity. Some people recycle tires by using them in gardening to surround and protect tomato and other plants.

Yard Wastes. Leaves, cut grass, and other yard wastes represent about 20 percent of all waste that ends up in landfills—about 35 billion tons a year. Yet these materials can be of use, and their recycling can save money. Yard wastes are usually recycled into compost, an organic substance that can be used to make soil richer for growing things. Composting takes place when yard wastes and other matter decompose under controlled conditions. Anyone can compost. You can buy simple composting devices from nurseries, gardener's supply companies, and from mail-order companies.

ECO-LABELING

It seems that for several years we have been awaiting the arrival of one or more systems to label products that are deemed to be "green," or at least greener than their alternatives. The two competing labeling organizations, Green Seal and Scientific Certification Systems (which was originally known as Green Cross), have been slow to certify products. And even when products receive one or the other group's blessing, they do not quickly show up on store shelves boasting an "eco-label."

Still, as this book goes to press, there are some encouraging signs that a new era of eco-labels is upon us. Both groups are increasingly announcing products that have met standards for environmentally responsible manufacturing or contents. And both groups have ambitious plans for the future. We have indicated just a few of those products in appropriate chapters in Part II of this book. You are encouraged to contact both groups to receive the most up-to-date lists of certified products, as well as other educational materials they may offer.

Here is where to contact them:

- **Green Seal,** 1250 23rd St. NW, Ste. 275, Washington, DC 20037; 202-331-7337.
- **Scientific Certification Systems,** 1611 Telegraph Ave., Ste. 1111, Oakland, CA 94612; 510-829-1415.

In seeking out eco-labels, keep in mind that many of the good, green choices to be found in supermarkets, department stores, hardware stores, and other places do not and probably will never carry an eco-label. For example, a product that is wrapped in far fewer layers of packaging than a competing brand will likely never be certified as "green." And products made from companies with better environmental records than competitng products also aren't likely to carry an eco-label for this reason alone.

So while both eco-labeling programs deserve the support of all Green Consumers, it is important to point out that no single labeling system will ever provide all the information we'll need to make environmentally wise choices. It will be up to concerned consumers to keep informed about what makes some products more environmentally responsible than others.

Part II

SHOPPING FOR A BETTER ENVIRONMENT

AUTOMOBILES

Cars—we can't live with them or without them. Take them away and some people's entire lives may need to be overhauled. Keep driving them and we may all be suffering from their many effects on the environment and on our lives.

Is there a happy medium? Is it possible to be a Green Consumer and still enjoy the luxury of independent mobility that our society treasures? Perhaps. Much of the answer depends on the kinds of cars we drive and on the way we care for them.

Before we explain, let's first take a look at the effects of cars on the environment. You probably already know about air pollution, especially if you live in a big city. It used to be that only a few cities—Los Angeles, Tokyo, and Mexico City, for example—were choking in car exhaust. Now, that list includes more than two dozen cities in the United States alone.

There's no question that automobile exhaust is one of the principal ingredients of environmental problems, including air pollution. In fact, cars and trucks can be implicated directly in most of the biggest environmental problems we face, from global warming to water pollution to everyday smog.

That's just the beginning of the environmental problems caused by cars and trucks. Consider the trash: Some 260 million tires, 80 million automobile batteries, and 400 million disposable oil filters are thrown away each year, each contributing its share of pollutants. And then there are the cars themselves. The 10 million or so cars sold in the U.S. each year consume nearly 9 billion tons of steel, plus vast quantities of other raw materials. Close to 9 million cars are junked each year, only a fraction of which are recycled as scrap metal. The rest fill our landfills or litter our landscape.

None of this includes the impact of oil exploration and drilling on the environment, and the oil spills that all too frequently result, wreaking havoc on every plant, animal, and organism for hundreds of miles around. Cars and light trucks guzzle more than a third of all oil used in the United States.

Why are cars so polluting? Haven't we enacted clean-air laws that have mandated automobile fuel-efficiency standards? Aren't cars getting more environmentally sound?

True, federal law has dictated certain fuel-efficiency requirements for two decades, since the enactment of the Clean Air Act of 1970. But a funny thing happened on the road to fuel efficiency and cleaner air: The number of cars on the road has surged more than 50 percent since 1970, to about 135 million. One problem is that from almost the day the clean-air law was enacted, American automobile manufacturers have fought the fuel-efficiency standards, often succeeding in weakening them. (Detroit's argument is to wait things out: The environment will clean itself up as more consumers replace their older, gas-guzzling cars with newer, more efficient ones. Clearly, automobile manufacturers have failed to learn the lesson of the past two decades: More cars create more pollution, no matter how efficient they may be.) As a result, even with more fuel-efficient cars, the nation's air is far dirtier than when the landmark law was passed.

A Greener Car?

It's not that car makers aren't capable of building a more fuel-efficient, less-polluting car. They need only follow the technology pioneered by the Japanese, such as multi-valve engines (which have four valves per cylinder instead of two), continuously variable transmissions (which lack gears, instead adjusting constantly to a car's changing speed), and "lean-burn" engines (which force a swirling layer of air and gas into each cylinder, enabling the engine to burn a "leaner" mixture—one that has a greater air-to-gas ratio, thereby producing the same amountof power with less fuel).

It is these and other features that have long allowed the Japanese to hold the number-one position in fuel economy for cars sold in the United States. (The Chevrolet Geo, since 1990 the most fuel-efficient car sold in this country—imports or domestics—which is rated at 53 miles per gallon in the city and 58 miles per gallon on the highway, is made by Suzuki Motor Co. and imported from Japan.) According to a report by the federal government's Office of Technology Assessment, such technology would help American auto makers achieve a 33-miles-per-gallon average—up considerably from today's 27.5 mpg—"without performance loss or the need to move to smaller vehicles." Additional costs of these innovations would be more than covered by increased fuel efficiency.

So despite the many controversies surrounding finding new sources of oil, it can be said that America's largest "oil reserves" have no environmental impact: They can be achieved by making fuel-efficiency improvements in American-built cars.

Here is a brief look at some of the most promising technologies:

• **Methanol:** Methanol's real benefits come from the fact that methanol is low in reactive hydrocarbons and toxic compounds. Moreover, methanol-fueled trucks and buses emit almost no particulates and far fewer nitrogen oxides than diesel-powered versions. Another benefit of methanol is that it can be manufactured from a variety of carbon-based feedstocks, such as natural gas, coal, and such biomass materials as wood. Finally, methanol is much less flammable than gasoline, and results in less severe fires when it does ignite. That could help reduce the half-million vehicle fires and 1,400 related fatalities each year in the United States.

One downside of methanol is that it is far more corrosive than gasoline, and could reduce the useful life of the fuel system, if not the car itself. Methanol-powered cars typically have stainless-steel fuel lines, which are considerably more expensive than other kinds of steel used in fuel systems. Another drawback of methanol is that it has less energy content per unit of volume than gas. So, while a 16-gallon tankful of gas might power a car close to 285 miles, the same-sized tank of methanol might be good for only about 200 miles. On the other hand, the price of a gallon of methanol is only about half that of gasoline.

All of the major auto manufacturers have produced cars that run on "M85," a blend of 85 percent methanol and 15 percent gasoline. Some cars are equipped to handle a varying gas-methanol blend. As fuel is pumped to the engine, it passes an optical sensor that measures the percentage of methanol by its light-bending properties. A computer alters the fuel's flow rates and ignition-timing to compensate for the changes.

• **Ethanol:** Produced from corn, barley, rye, wheat, sugar cane, sugar beet, and other crops, this fuel already powers 2 million vehicles in Brazil—most of them manufactured by General Motors and Ford. According to Columbia University professor Harry P. Gregor, "If ethanol use were mandated for all city-owned or regulated vehicles such as buses, taxis, police cars, and garbage trucks in the nine U.S. areas of highest air pollution, air quality would rapidly and markedly improve." The cost of changing to ethanol was tremendous, however, and initially led to worsening the air pollution problem in Brazil, due to misadjustments in engine conversions that created high evaporations. A new problem, especially in northeastern Brazil, arose when sugar prices skyrocketed and the price of sugar beets rose as well, producing a shortage of ethanol.

In the United States, "gasohol"—gasoline containing about 10 percent ethanol—can be found at select service stations, particularly in the Midwest, where its popularity stems from the fact that ethanol can be made from corn.

• **Flexible-fuel engines:** Many experts predict that when alternative-fuel cars do come out—probably toward the end of the decade—they will be able to use either methanol or gasoline, permitting motorists to opt for the most available and affordable fuel at any given moment. In 1986, both Ford and General Motors introduced special models that can run on gas, methanol, and several other fuels—singly or in combination. But neither Ford's "flexible fuel" Crown Victoria nor G.M.'s "variable fuel" Chevy Corsica is yet being mass-produced.

• **Natural gas:** This represents one of the best prospects for fueling trucks. Compressed natural gas (CNG), as it's formally known, emits about 80 percent fewer hydrocarbons, 90 percent less carbon monoxide, and 22 percent fewer nitrous oxides than does gasoline. And there's plenty of natural gas available from domestic sources. The Energy Department estimates that existing pipelines could supply enough to replace 1 million barrels per day of gasoline and diesel fuel and run 8 million vehicles.

Currently, about 30,000 vehicles powered by natural gas are in use around the country, according to the Natural Gas Vehicle Coalition. Natural gas powers buses in Alaska and Florida. United

How Much Pollution?

(Grams per passenger-mile for typical work commutes)

Mode	Hydrocarbons	Carbon Monoxide	Nitrogen Oxides
Rail	0.01	0.02	0.47
Bus	0.20	3.05	1.54
Vanpool	0.36	2.42	0.38
Carpool	0.70	5.02	0.69
Single-person auto	2.09	15.09	2.06

Source: American Public Transit Association

Parcel Service, the world's largest package delivery company, is undergoing an experiment that eventually could result in a large portion of the company's 104,000 vehicles switching to natural gas. So are Federal Express and several other delivery companies. Other company fleets use some natural gas-powered vehicles, too.

Retrofitting vehicles to run on natural gas is costly—around $2,000 per vehicle, and only a handful of shops around the country are capable of making the switch. But the installation price is offset by the lower price of natural gas—around 75 cents for the energy equivalent of a gallon of gasoline. The 50-cent difference between CNG and gasoline would enable the gas retrofit to pay for itself after the equivalent of 4,000 gallons—5 to 7 years' worth for a typical car driven 15,000 miles a year. Another problem is availability of CNG: While the gas itself is plentiful, there are only about 125 CNG dispensing stations open to the public in the United States. And then there's the problem of cruising range. The CNG tanks now available hold only the equivalent of about five gallons of gasoline. That means far more fill-ups.

Each of these alternatives offers distinct advantages and disadvantages. And none of these—or any other—alternatives will work without an abundance of these fuels as close as the corner service station. Indeed, the lack of widespread availability of these fuels may be the biggest obstacle of all for alternative fuels. Oil company resistance to having to produce and market alternative fuels along with gasoline and diesel has slowed progress considerably.

• **Reformulated Gasoline:** The oil companies would rather that Americans stick with gasoline—albeit a cleaner, more efficient gasoline. Indeed, much of the antipollution efforts of several oil companies have focused on reformulated gasoline. ARCO, for example, has hyped its new "EC-1 Regular" (the "EC" stands for "emission control"), which it introduced in southern California in 1989. Intended for the heaviest-polluting vehicles—pre-1975 cars and pre-1980 trucks that run on leaded gas and do not have emissions-reducing catalytic converters—the fuel was claimed by ARCO to cut automobile pollution substantially, without affecting performance or costing consumers extra. Actual evidence is inconclusive. According to ARCO, "If all leaded gasoline users in southern California switched to a gasoline like ARCO EC-1, about 350 tons of pollutants would be removed from the air each day. The impact would be equivalent to permanently removing 20 percent—or more than 300,000—of these high-polluting, pre-catalyst-equipped vehicles from Southern California highways."

Another ARCO gas, dubbed "EC-X" and announced in 1991, is said to burn as clean as methanol, cutting some pollutants by one-third. In its announcement, ARCO modestly called its product "the cleanest-burning gasoline ever developed." But ARCO tempered its remarks by noting that it would not market EC-X, which costs 15 to 20 cents more per gallon to produce, unless it was ordered to do so by air-quality regulators.

Be careful: Reformulated gas isn't for all cars. One authoritative study found that in 1983 through 1985 model cars, reformulated gas actually increases emissions by about 9 percent. No one knows exactly why. It's one of those many unanswered riddles.

Waiting for Cleaner Fuels

Ultimately, reformulated gas is a dead-end. The oil companies will have to give in to the realities that the least polluting fuels are non-petroleum-based, such as natural gas and electricity. There is no known technology using gasoline that will reduce automobile pollution to acceptable levels. While improvements will continue to be made in both automobiles and their fuels, the time is not far away when a sizable portion of vehicles will operate on alternative fuels. Someday, all of them will.

That day would arrive sooner if the federal government would strengthen its clean air rules. Meanwhile, the debate between car-makers and law-makers centers around whether the average mileage for all cars produced by a given manufacturer should be 26.5 or 35 or even 40 miles per gallon. But those arguments obscure the real point: To reduce the levels of pollution in some areas to acceptable levels will require cars that run on new, cleaner fuels.

Even while we wait for these fuels, there's a lot that can be done with the gas-powered car. Most experts believe that the technology exists to raise the average to 35 or even 45 mpg. If the average car on the road achieved that level of efficiency, the United States would have to import almost no foreign oil. But car makers have said that raising the average mileage to this level is not possible.

But their arguments were given less credence in 1991, when Honda Motor Co. and Mitsubishi Motors Corp. unveiled new "lean-burn" engines said to be capable of powering a car up to 60 miles on a gallon of gas. Experts quickly pointed to the engines as evidence that the technology does already exist to build more fuel-efficient cars. Honda introduced its new VTEC-E engine in some 1992 Civic VX models. It also has exhibited a prototype car that gets 100 mpg. Several other car makers have built prototypes with impressive fuel-efficiency—Volvo's LCP 2000 gets 65 mpg, Volkswagen's E80 gets 74 mpg in the city and 99 mpg on the highway, and Toyota's AXV gets a whopping 89 mpg in the city and 110 mpg on the highway. Even Ford has an unnamed prototype that gets 57 mpg city, 92 mpg highway. Other cars have gone as far as 138 miles on a gallon of gas.

But, alas, such cars always seem to be a few years away from being available in America's automobile showrooms.

THE ELUSIVE ELECTRIC CAR

What about cars powered by electricity? It's certainly an attractive proposition: Just plug it in, charge it up, and take off. The fact is, it's already possible to do just that. Electric cars have been on the market for a few years.

Why don't you—or anybody you know—drive an electric vehicle (except, perhaps, on the golf course)? Because they're being manufactured only by a small number of small companies, who make only a small number of cars each. The sum total of their output isn't enough to meet the slowly but surely growing demand for these virtually pollution-free vehicles. As a result, some electric car makers have production backlogs two or three years long.

What makes electric cars so attractive? Here are several reasons:

- Electric cars are virtually maintenance-free. And with good reason: they have no spark plugs, valves, mufflers, air or fuel filters, hoses, fan belts, water or fuel pumps, pistons, radiators, condensers, points, starters, or catalytic converters. As a result of all this simplicity, they need no tune-ups, oil changes, or radiator flushes, among other things.
- They are virtually pollution-free, too. They give off no exhaust and make little noise. Even including the power plant emissions that result from generating the electricity needed to run the cars, they still emit 90 percent fewer pollutants, according to the California Air Resources Board.
- They don't need to be warmed up. Whatever the temperature outside, you can simply turn the key and get going without worrying about stalling out.
- They can save money. At current electric rates, you can drive an electric car for about 3 cents per mile. A 25-mpg car costs about 6 to 8 cents in gas and oil, not including higher maintenance costs.

But electric cars also have their drawbacks. The biggest has to do with their ability to hold enough electricity to travel more than 50 to 75 miles before needing to be recharged. A related problem is the time it takes to fully recharge an electric vehicle—four to eight hours. Both problems have to do with limits to battery technology. Until now, we never needed batteries to hold so much power for so long. But even these limits don't rule out electric cars when you consider that most people drive their cars only a few miles a day anyway. So, electric vehicles can make sense for a daily commute of 10 to 15 miles each way.

Meanwhile, battery technology is changing quickly, as more companies pour research dollars into the effort. In 1991, the Big Three American manufacturers formed a consortium with the federal government, utilities, and battery makers to develop batteries with greater range and shorter recharging times. That same year, Nissan Motor Company said it had developed new batteries and a system that could reduce recharging time by 80 percent. Nissan says its system can be recharged to full capacity in 12 minutes—not much longer than it takes to fill your tank.

How about all the electricity that will be needed to power the cars? Would the increased demand lead to more nuclear power plants, coal-fired generators, and other environmentally undesirable components of electric utilities? No, according to a report by the Office of Technology Assessment, a branch of Congress. The United States already has enough surplus electric-generating capacity to run "several tens of millions" of electric vehicles if they are recharged at night, when other electricity demand is low: Recharging the latest generation of cars takes about as much power as a color television uses in about six hours.

Still, it will be a few years before electric cars are available from any of the major manufacturers. General Motors, for example, gave up on its plans to be first off the block with its curiously named Impact, opting instead to put its efforts in the Big Three consortium The Impact—which had been described in company literature as "environmentally friendly, aerodynamically superior, eminently practical, and aesthetically exciting"—would have reduced some emissions by 96 percent compared with gas-powered equivalents. According to one newspaper reporter, the Impact was no golf cart. It accelerated from 0 to 60 miles per hour in about 8 seconds, "enough to beat the Mazda Miata and Nissan 300ZX." In the end, the Impact never made it into full-scale production. It's something of a chicken-and-egg situation: Manufacturers won't make the cars unless there's a demand, but consumers won't demand them until they're readily available.

In the end, the burden will lie with car buyers. If we demand electric vehicles, the companies will make them.

Sources of Information on Electric Cars

There is a great deal of information on electric vehicles. Here are some manufacturers that can provide more information. Write for their current catalogs and spec sheets.

• **General Motors Corporation** (432 N. Saginaw St., Ste. 801, Flint, MI 48502; 800-253-5328) will provide information about its Impact model.

• **Solar Car Corporation** (1300 Lake Washington Rd., Melbourne, FL 32935; 407-254-4566) is a small company working to develop solar-electric and alternative-fueled vehicles. Its Solar Festiva seats four, has a top speed of 60 mph, and recharges overnight. A hybrid model with a small internal-combustion engine is also available.

• **Solar Electric Engineering, Inc.** (116 4th St., Santa Rosa, CA 95401; 707-542-1990) offers the Electron 1, a refitted Ford Escort hatchback; and the Electron 2, which uses a plastic-body, two-seat Pontiac Fiero. The Electron 1 has a range of up to 100 miles; the lighter Electron 2 goes slightly faster and farther. The company also offers a $6,000 kit to turn almost any car into an electric one.

BUYING GREEN

After two decades of fighting for the right of every citizen to drive the biggest gas-guzzler he or she can afford, has the automobile industry suddenly seen the light?

To find the answer, take a stroll into the auto showroom of your choice. Ask the friendly salesperson to show you the environmental features of this year's crop of cars. Chances are, you'll be directed to the cars' EPA gas-mileage ratings and not much further.

So what's all the fuss about?

In answering that, let's start with the good news: Nearly every car maker is working feverishly to make environmental improvements in its products. The bad news is that most of these improvements are several years away from mass production, and most cars won't sport more than a few of the improvements—at least for a while. Most of the technologies that will make cars drive cleaner are still in the experimental stage. And the feature that would make the single biggest difference to the health of the planet—higher fuel-efficiency standards—is still being fought bitterly by car makers.

All of which is to say that the environmentally correct car is miles down the road. Still, the types of features hyped at the Automotive

News conclave weren't just pipe dreams. Some good, green innovations have arrived.

But the reality is this: There is no such thing as a perfectly "green" car. In fact, there's nothing even close. (Indeed, you'll scarcely even find a car that's painted green: According to a 1990 study by DuPont, less than .5 percent of all sports cars in America are green, In contrast, more than 30 percent are red. But that's another matter.) So, the search for a lean, green machine can be frustrating. But it needn't be futile.

What makes a car environmentally responsible? A lot has to do with how you drive it, maintain it, and dispose of it, issues we'll cover in subsequent chapters. But first, let's look at some ideal criteria for "greenness":

- An environmentally responsible car should be as fuel-efficient as possible, getting the highest possible miles per gallon among cars of its size and class.
- It should include components that enhance fuel use, such as radial tires and fuel-injection.
- It should be durable, both in its construction and in its features. Rather than have disposable parts, such as fuel filters and spark plugs, it should wherever possible use longer-lasting products.
- It should, wherever possible, use nontoxic or nonpolluting fluids and systems, such as non-CFC air conditioning coolants and nontoxic antifreeze.
- It should be manufactured using products and processes that represent the cutting edge of environmental technology, such as low-emissions paints and finishes.
- It should be made by a company with a good environmental record, relative to other auto makers.
- Ideally, it shouldn't cost more than its "nongreen" counterpart.

Granted, this is a tall order. For one thing, it is still virtually impossible to find a car that meets all of these requirements, and some of them—such as manufacture by an environmentally responsible automobile company—are not yet within reach, not by a long shot. Some of the criteria mentioned, such as the use of low-emissions paint and finishes, are still in the experimental stages. As for the last criteria—price—it's elusive at best, given the slippery nature of automobile pricing. Variations can include base price, dealer prep, trade-ins, options, and haggling, along with other sordid elements of this generally unpleasant negotiating process.

So, finding the "greenest" car possible remains largely a hit-or-miss proposition. In choosing the size of car you want, you may or may not find the options and features on it that could give the car an environmental edge over other models.

Buying Tips

Still, there is some basis on which to compare cars. Let's take a look at some of the considerations of car buying, and how they can affect the environment. Here are some simple suggestions.

• **Get the most fuel-efficient car you can buy.** This scarcely even needs to be said, but let's get it out of the way. The more miles your car gets from every gallon of gas, the less damage it will do to the environment, and the less it will cost you to own and operate.

• **Don't buy more car than you need.** If you're mostly going to be tooling around town by yourself, perhaps hauling a few grocery bags and a friend or two, you don't need a huge vehicle. Look for a smaller model with a smaller engine. Otherwise, you'll be wasting your money.

• **Don't get a lot of power gadgets.** Let's face it, rolling down the windows by hand isn't such a big deal. Locking the doors by hand is pretty straightforward. And moving your seat back and forth can be done just as easily with a mechanical lever than with an electronic switch. The fact is, many of these options not only cut fuel economy, they are also costly to repair when they break. And they inevitably do.

• **Opt for cruise control.** The above rule notwithstanding, there are a select few options worth considering. Cruise control is one of them. This is a simple set of controls, usually located within the steering wheel, that allow you to take your foot off the gas and let the car take over. Computers inside the engine automatically regulate the gas and brake to maintain whatever speed you set. (Obviously, this is only useful if you do a a lot of highway driving. It's not very helpful in stop-and-go city traffic.) Computers regulate speed a lot more efficiently than humans do, resulting in better gas mileage. It also makes long-distance driving considerably more

The Ecology of Used Cars

When you examine the evidence, there is a good case to be made that if you must get another car, buying used is the greenest choice. After all, "reuse" is the second-best choice among a green consumer's "Three Rs." (The first choice would be "reduce"—that is, not having a car at all.) Buying a used car makes sense if the vehicle you buy gets better gas mileage—and is therefore less polluting—than your present car.

When you purchase a used car, you avoid putting yet another polluting vehicle on the road. But you also can get more car for your money. When you buy a new car, you pay for a lot more than just the vehicle you drive away. For each dollar you spend on a new car, you spend about two cents for transporting the car from the factory to the dealer's showroom; eleven cents for dealer preparation, markup, advertising, and other forms of profit; six cents for registration and sales and other taxes; and just eighty-one cents for the actual vehicle.

In other words, if you pay $12,000 for a new car, only about $9,720 goes for the car itself; the other $2,280 goes to help the dealer and manufacturer make the sale.

Well, $2,280 to buy a little peace of mind and status may not seem all that much, but that's just the beginning. The minute you drive your new car out the dealer's door, its value goes down by another 20 to 25 percent—between $2,400 and $3,000 for that $12,000 car.

The bottom line: you spend $12,000 to get a little more than $7,000 worth of car.

True, used cars have some profit, too, especially when buying from a dealer. (Studies have shown that the best car buys are from private individuals.) And unless you are a skilled mechanic, you *must* have a used car inspected by an independent mechanic before you buy it. But if you want to get the most car for your dollar, buying used may be the way to go.

What are the best buys in used cars? Check with reliable sources, such as *Consumer Reports,* which rates used car models every year in its April issue, or consult *The Used Car Book,* by Jack Gillis, an annual guide published by HarperCollins.

comfortable. This option, if it's available, is worth considering.

• **Avoid power boosting options.** Unless you plan to be drag racing down the interstate, you simply don't need turbochargers. They're expensive (typically around $1,000), they break down, and they waste gas. Besides, they're designed for race cars, not for stop-and-go driving. They'll cost you money from the day you buy the car until the day you unload it on some unsuspecting fool.

• **Consider color.** You probably don't consider a car's color an environmental consideration, but it is. Simply put, don't buy a dark car if you live in a warm climate, even if it gets cold in the winter. As you may know, dark-colored cars absorb heat, whereas lighter colors reflect heat. So, in summer, darker cars will get hotter inside, requiring more air conditioning to cool you down. That, of course, is a big drain on the fuel efficiency. Of course, in winter, a lighter car will be colder, but heating a car takes less energy than cooling it. The reason: When you turn on the heater, you simply direct some of the engine's heat into the passenger area. The heat is "free" in that it exists whether you use it or not. But air conditioning requires turning on energy-intensive compressors as well as fans. So, opt for lighter colors. By the way, that goes for your car's interiors, too: Darker fabrics stay hotter longer, requiring the air conditioner to work even harder to cool things down.

• **Avoid vans and light trucks.** According to the California Energy Commission, cars emit 70 percent less carbon monoxide, 50 percent fewer hydrocarbons, and 20 percent less nitrogen oxides than vans and light trucks, while using 30 percent less gasoline over the same number of miles.

DRIVING GREEN

The fact is, most of us drive pretty inefficiently. We don't do it on purpose. A lot of it has to do with our lifestyles. Statistics show that the average automobile trip in the United States is about 9 miles. And it's during those very same miles—the first few minutes of driving—that your car uses a lot of gas and creates a lot of pollution.

Why is this? When your car's engine is cold, it requires a higher ratio of gasoline to air to operate. More gas equals more air pollution. In old cars, there was a gadget on the dashboard called the choke. You usually pulled it out when starting the car to increase the gas-to-air ratio. Now, most cars have automatic chokes. You may not be aware that the process is taking place, but it is.

Another problem is that gasoline needs to be heated up and turned into vapor to work in the engine. Needless to say, things don't heat up as easily in a cold engine and a surprising amount of the gas never gets vaporized. It simply trails out the tailpipe in its original liquid form. Of course, that creates even more pollution.

Other engine parts also need to be warmed up to work efficiently. Consider catalytic converters. These are remarkable little devices, the most successful technology ever invented to curb air pollution. In fact, today's catalytic converters eliminate 96 to 98 percent of carbon monoxide and hydrocarbon, and three-quarters of the nitrogen oxide emissions.

How do they do this? Catalytic converters work by burning up pollutants, converting them into three basic elements: carbon dioxide, water, and nitrogen. That's all there is to them.

The bad news is that catalytic converters need to be warmed up to do their work.So until a car's engine is warmed up, the burning process is incomplete. Indeed, when you first start a cold engine, the converter is too cold to burn anything. Virtually all the pollutants are emitted through the tailpipe.

Here's the bottom line: During a typical 20-mile commute, half of the hydrocarbons will be emitted during the first three or four miles. Reducing cold starts and short trips are the two most important things you can do to increase gas mileage and decrease pollution.

Engineers are working on some high-tech solutions that will help catalytic converters warm up faster, perhaps instantly, so they do their good deeds from virtually the first minute you turn the key. But until that happens, it's important to know the environmental hazards of a cold engine.

Getting Off to a Cold Start

How do you avoid these problems? Here are four suggestions:

- **Learn the proper way to start your engine.** This may sound rather simplistic, but you'd be surprised how much fuel is wasted by people needlessly pumping the gas pedal when starting a car. In fact, if your car has fuel injection, you usually don't need to press the pedal at all when starting. If your car is in tune, it should start right up, even when it's cold outside. Cars with carburetors may require pumping the gas once, maybe twice. This is one place where your owner's manual will come in handy. Few individuals bother to read them—or at least the part that describes proper starting procedures. Pull yours out and spend two minutes going over it. You might learn a thing or two.
- **Don't warm up your car by idling.** Once upon a time,

commuters used to sit in their cars, turn on their engines, and read the paper for five or ten minutes before taking off. This, they believed, would warm up the engine so as to minimize wear and tear. We now recognize that the opposite is true: Idling a cold car will increase engine wear and tailpipe emissions. If your car has been properly tuned and is otherwise in good shape, you needn't idle your car more than a few seconds before taking off. Those first few minutes of driving will do more to heat up the engine—and get the catalytic converter working—than anything else.

• **Start driving slowly.** You shouldn't go from 0 to 60 during those first few minutes. Take it easy. Accelerate gently. Try not to go over 35 miles per hour for the first minute or so.

• **Don't turn on the heat right away.** This may be difficult to do on a bone-chilling morning, but doing so will rob your engine of precious heat during those first few crucial minutes. Besides, all you'll get during the first couple of minutes is a blast of cold air. Try to wait three or four minutes before turning up the temperature.

If you live in an area with extremely cold winter weather, you might consider investing in an engine heater. These handy little devices keep the oil and other components "preheated" overnight, so your car is ready to drive in the morning. While it may sound like a waste of energy to keep your car warm at night, it's not. The energy consumed is rather low and is further offset by the substantially reduced driving emissions when you start a prewarmed car. Some newer cars actually have outlets built into the engines where you can install an engine heater—one from the auto manufacturer, of course. But there are many "after-market" brands available, costing as little as $20. They're worth it.

Green Driving Tips

Here are more do's and don't's for driving environmentally:

• **Observe speed limits.** The typical car uses 17 percent more gas when driven at 65 miles per hour than at 55 mph. At 70 mph, you use about 25 percent more gas. On the other hand, don't drive too slow. Most cars get their best gas mileage somewhere between 35 mph and 45 mph.

• **Avoid jump starts.** Never put the "pedal to the metal"—that

Keeping Your Car Running Green

Here are some tips on keeping your car running longer—and polluting less.

• **Check your tires.** Simply keeping your tires inflated can save gas and tire wear. Underinflated tires can decrease fuel economy by as much as 5%.

• **Use the right oil.** Consider using synthetic oil products, such as Mobil 1, which reduces engine wear and increases the time between oil changes. However, be aware that synthetic oil is not yet recyclable, so it cannot be mixed with regular oil.

• **Don't skimp on oil filters.** Always buy top-of-the-line brands such as Purolator and Frantz. Cheaper filters don't screen oil as thoroughly and can clog over time. When that happens, the oil goes through the bypass valve—sending unfiltered oil into the engine, which could harm it.

• **Go permanent.** Even better, consider buying a permanent oil filter. One such product is the System One filter (System One Filtration, P.O. Box 1097, Tulare, CA 93275; 209-687-1955). It costs about $100, but it's supposed to last for up to 100 oil changes, a considerable savings over the $5 or so you pay for a disposable filter. A permanent oil filter also helps cut the number of filters ending up in landfills. Almost 400 million filters are disposed of each year, each of which may contain up to a quart of oil. Cleanable and reusable air filters are also available.

• **Go platinum.** Platinum spark plugs, which cost 2 to 3 times as much as conventional plugs, last up to 100,000 miles—far longer than regular plugs do. That means you'll need to change them less, which will save money in reduced maintenance costs. As plugs wear out, they affect the engine's timing, reducing gas mileage. Because platinum plugs wear out much more slowly, they can boost fuel economy.

• **Buy parts for a lifetime.** Ask your mechanic to buy parts that have a lifetime warranty, such as brake parts, water pumps, alternators, and starters. They usually don't cost much more than other parts. While they may not actually last a lifetime, these parts are usually better built and manufacturers will usually replace them if they break.

• **Read your owner's manual.** You may have seen it buried somewhere in your glove compartment or at the bottom of a clutter drawer. Owner's manuals contain a gold-mine of information on how to keep your car running at peak performance.

is, floor the accelerator—unless your life is in danger. Doing so can burn as much as 50 percent more gas than a relatively smooth start. Among other things, flooring the accelerator sends more gas than the catalytic converter is able to deal with. The result: Wasted, unburned gas is emitted through the tailpipe—and into the air.

• **Accelerate smoothly and moderately.** When you get to your desired speed, keep steady pressure on the accelerator, just enough to maintain your speed. Speeding up and slowing down wastes gas. Here's a trick: Pretend there's a full glass of water sitting on the seat next to you. Your goal is to get the car moving at cruising speed without spilling any of it.

• **Ride the "bubbles."** Traffic often forms patterns in which there are clusters of cars interspersed with "bubbles" of few or no cars. When possible, position yourself in one of those "bubbles" so you can keep a steady speed. It's much safer riding in bubbles, too.

• **Lose some weight.** In your car, that is. Remove unneeded stuff from your trunk—golf clubs, books, cans of oil, whatever. The lighter the car, the less gas it uses. An extra 100 pounds decreases fuel economy by about 1 percent for the average car, slightly more for smaller cars.

• **Use cruise control.** If you have this option, use it on the highway to help even your speed and reduce your gas use. But don't use it in town. Cruise control is ineffective in stop-and-go driving and can actually decrease your gas mileage.

• **Don't ride your brake.** You'd be surprised at how many people drive around with one foot on the accelerator and the other on the brake. This is unsafe, both for you and the car. It wastes gas and prematurely wears down brake pads and shoes. When you see a red light or stop sign ahead, or otherwise know you'll be stopping the car, take your foot off the gas well in advance. That will save gas by giving your car a chance to slow down by itself before applying the brake.

• **Don't idle.** When your car is idling, it is getting zero miles per gallon! If you must stay in one place for more than about a minute, it makes more sense to turn the engine off. It takes less gas to restart the car than to let it idle. Don't forget this rule when you get in the car to go somewhere. A tremendous amount of gas is wasted after people start the engine, then read a map, rearrange the front seat, eat a sandwich, or whatever.

• **Don't tailgate.** Following too closely requires much more braking and accelerating.

• **Plan your trips.** Select routes that will allow you to consolidate errands, avoid congested areas, and avoid needless driving.

• **Turn off your air conditioner.** Driving with the air condi-

Shop Talk

What makes an auto repair shop "green"? There's no single definition, but there are some clear-cut signs:

• **Recycling of antifreeze.** Using a recycling device, mechanics can capture used antifreeze, which can then be filtered to remove acids and sediments before being recycled back into car radiators. The recycled antifreeze, according to some experts, is at least as good as the new stuff.

• **Recycling coolant.** A similar technique is used with Freon, the gas used in most car air conditioners. Enlightened mechanics use a "vampire," a device that captures the Freon and saves it from doing its environmental damage. Like antifreeze, Freon can then be cleansed and recycled in other cars' air-conditioning systems.

• **Recycling motor oil.** Because it is illegal to dispose of used oil down the drain, virtually all shops save their oil for recycling. (Used oil can be recycled into other kinds of lubricants, including fuel oil.) The truly responsible shops allow customers who change their own oil to bring the used oil to the shop for proper disposal.

• **Buying oil in bulk.** Some shops still buy oil by the quart, just like those sold in supermarkets and service stations. Most cars use three to four quarts of oil, so at the end of a workday, an auto shop can amass hundreds of cans—a small mountain of trash, to say the least. There's no reason shops can't buy oil in 55-gallon drums. For one thing, they'll save money. For another, the drums can be refilled.

For more information contact the **Sustainable Transportation Repair Society** (1350 Beverly Blvd., Ste. 115-324, McLean, VA 22101; 703-893-1239), a membership organization for auto mechanics to promote environmentally sound vehicle repair practices.

tioner on can waste up to a gallon of gas per tankful, especially in city traffic. Instead of automatically turning the air conditioning on, try the vents, blowers, and windows first. That may do the trick.

• **Learn how to shift.** If you have a manual transmission vehicle, proper shifting involves more than getting the car from one gear to the next smoothly. The goal is to get into high gear as quickly as possible without straining your engine. In some cases, you may even skip gears—going from second gear directly to fourth gear, for example. But don't go too high too fast or the engine will "lug," with wasted gas pouring into the engine. Lugging also is common when a car is slowing down. You can prevent it by shifting into a lower gear when decelerating. In a car with automatic transmission, accelerating gently helps you get into higher gears more quickly. It may also help if you "let up" a little on the accelerator.

• **Maintain momentum.** It takes energy to accelerate, of course, so the less you have to brake, the less you'll have to accelerate to regain speed. It takes 5 to 6 times as much gas to get a car rolling from a dead stop than when it's already rolling, even at a few miles per hour.

• **Avoid peak-period travel.** If you can commute or otherwise travel during non "rush-hour" times, you'll save a great deal of gas—and time. When average speeds drop from 30 to 10 mph, fuel consumption doubles.

A FUEL'S PARADISE

The simple act of putting gas in your car can make a difference in your car's performance and fuel efficiency. There are three basic considerations: the octane level of the gas you use, the kind of gas you buy, and how you fill your tank. Let's take a look at each.

Octane

Those dreamy television commercials run by gas companies—with dolphins jumping out of the water, bald eagles soaring high above, or fragile deer grazing in the forest—might lead one to believe that Chevron, Exxon, Mobil, Shell, Texaco, Amoco, and all the other oil companies are principally in the business of saving the earth from environmental peril. Of course, that's not really the case: They're in it to sell us gasoline and other products. And like any corporation aiming to stay in business, their aim is to sell you the greatest amount of the highest-priced gas they can get away with.

But most gasolines are essentially the same. Their high-tech sounding names are simply marketers' attempts to make something as plain and boring as motor vehicle fuel sound smart and sexy. Sure, each company pours in additives it claims gives its product superior performance, or offers benefits for your car—and they might even offer tests that prove it—and some of these may even

High Octane, High Price

According to a 1992 report from Public Citizen's Buyers Up, a Ralph Nader organization, "Motorists are being taken for a ride—an expensive one costing billions of dollars a year." First, oil company ads mislead consumers into buying gasoline that is higher octane than their engines require. Moreover, mislabeled pumps tout a higher octane gas than they really have—for which station owners still charge a premium price.

The report, titled *Fueling the Public*, specifically criticizes the two government agencies authorized to protect consumers on these issues, the Environmental Protection Agency and the Federal Trade Commission. The EPA is supposed to verify octane labels on pumps, which Buyers Up claims they haven't done since 1981. The FTC is supposed to take action against companies misleading the public; the first such enforcement action did not occur until this year, and the agency is yet to collect any money. "We are asking both agencies to better monitor the tactics of gasoline marketers in this country," says Buyers Up director Linda Beaver.

In the mean time, check your owner's manual to find out the proper octane gas for your car. The fact is, fewer than one out of ten cars manufactured since 1982 requires high-octane unleaded gas. Buying a higher-octane gas than you need is costly, and it also contributes to air pollution.

To obtain a copy of *Fueling the Public* ($10 for individuals and nonprofits; $35 for others), contact Public Citizen Publications, 2000 P St. NW, Ste. 605, Washington, DC 20036; 202-833-3000.

work. But brand shouldn't be your first consideration.

Instead, you need to look at octane.

Do you buy high-octane gasoline? About 40 percent of drivers do, though fewer than one out of ten cars manufactured since 1982 require high-octane unleaded gas. Buying all that unneeded octane is costly, to be sure, but it is also a major source of air pollutants.

The problem has to do with "aromatics," a category of hydrocarbons that boosts octane levels. Among the most common aromatics are benzene, toluene, and xylene, all of which are air pollutants.

Consider benzene. According to information submitted to the federal government by Chevron, "Repeated or prolonged breathing of benzene vapors has been associated with the development of chromosomal damage in experimental animals and various blood diseases in humans ranging from aplastic anemia to leukemia (a form of cancer). All of these diseases can be fatal."

Chevron's gas contains up to 4.9 percent benzene. Most other manufacturers' gas have similar levels. According to EPA, cars and other motor vehicles account for more than 80 percent of all airborne benzene.

Moreover, there is evidence that aromatics in gasoline help to foul car engines, decreasing fuel efficiency. Ironically, some motorists turn to higher-octane fuel to remedy the problem.

So, why do so many motorists opt for high-octane fuel? The demand stems from the millions spent by oil companies to push high-octane gas. At up to 50 cents more per gallon compared to regular gas, there is great incentive for oil companies to get you to switch (see box, page 75). The ads are misleading at best. They imply that octane has something to do with fuel quality, or with the power it provides. In reality, it has nothing to do with either. Octane is a measure of a fuel's ability to resist knocking—the premature ignition of the gas-and-air mixture in a car's engine. Simply put, the higher the octane number, the better the resistance to knocking.

Most cars don't need this extra octane. The lowest-grade gasolines do just fine. Only cars with high-compression engines, such as BMWs, Jaguars, and Porsches, require the highest octane.

The right octane level for your car is indicated in its owner's manual. Always buy the lowest octane gas that works for your car. Anything more will waste your money and pollute the air.

The Right Brand

Having already said that most brands of gas are basically the same, let's briefly reconsider. Most experts suggest sticking to the major brands, the ones whose advertisements you see just about everywhere. This suggestion is not intended to support Big Oil (or knock the little guys). But for your car's sake, there's nothing to be gained from buying from the smaller producers, especially when their product is essentially the same price as the big guys. One reason is that the smaller refiners typically do less business, so gas may sit around longer. That gives it the chance to collect impurities, which could impair your engine's performance.

Beyond that, as stated earlier, avoid super-premium gas and buy the lowest octane allowed for your car.

You may want to consider switching to gasohol. This, if you haven't heard, is a mixture of gasoline and alcohol, usually derived

Gas Gizmos

Getting more miles per gallon of gas has always been the holy grail for both drivers and environmentalists. As is the American way, entrepreneurs have offered a dazzling array of high-tech (and some pretty low-tech) gizmos, gadgets, and potions intended to boost plain old gasoline into some kind of superfuel, and your car into a lean, green machine.

Some of these things may work, but a lot of the "green" benefits seem to be found only in manufacturers' bank accounts.

Take the Vitalizer, a device that, when installed in a car's fuel line, "causes an interruption in the natural flow pattern of the fuel, activating an electrostatic charge within the fuel matrix, and thus forming an electrostatic colloidal matrix," according to one catalog. The bottom line is that the device is supposed to ensure "more complete combustion of the fuel utilized," boosting efficiency by 20 percent or more—and, of course, cutting emissions. It also claims to clean spark plugs, improve starts, and reduce carbon deposits.

Does it work? Depends on who you ask. One person will swear by the thing, another will swear at it. And both parties may be absolutely right.

It's not just car owners that seem to have questions about fuel enhancers. In 1991, the Federal Trade Commission warned consumers to be skeptical, too. And EPA tested dozens of these products and found only a few that worked, and none that significantly boosted mileage or reduced emissions.

The Vitalizer is just one of many such products. There's the Magnetizer, the Softron, the Magna Flo, and the Empower+. Each offers some remedy to unburned fuel, carbon buildup, or some other obstacle to high mileage. Prices are typically $50 to $150, plus installation.

While none of these products has been exposed as outright fraud, federal authorities have begun to crack down on some of the more outrageous claims. The FTC sued the maker of the PetroMizer for claiming, among other things, that it increases gas mileage by 28 percent, reduces pollution, was developed under a grant from NASA, and was tested favorably by the Army—none of which apparently is true. At least one state attorney general has sued the Magnetizer Group to stop it from selling the Magnetizer.

In the end, there are no miracle cures. The quest for fuel efficiency is decidedly low-tech. Success comes from keeping your car in shape and driving it right. These things alone will go a long way to making sure you won't waste your fuel—or your money.

from corn. It is available with varying reliability. It is most prevalent in the Midwest, where it is popular because it benefits the farming industry. But it's also found in other places.

Gasohol isn't for everyone. It is not recommended for cars that aren't already driving smoothly and not experiencing problems. In fact, some mechanics believe that gasohol has a tendency to worsen existing problems. But if your car is generally well-maintained and in good running condition, gasohol should work just fine.

Filling Your Tank

"Fill 'er up," has become as much a part of our language as "How do you do?" But some people take the phrase a bit too literally. When filling their vehicles with gas, they're not satisfied when the gas pump automatically clicks off, indicating a full tank. They want more. So they "top off" their tank, pumping in another half gallon or so until the gas tank literally brims over.

These folks may think they're doing themselves a favor with this extra pumping. Perhaps they'll get to drive an extra 15 or 20 miles before having to refuel. In fact, they're doing everyone a disfavor. The reason: Topping off causes pollution. That's because when gas is heated as you drive, it expands in the gas tank. If the tank is already filled to the brim, some gas spills out, contributing hazardous fumes into the atmosphere.

But you needn't overfill your tank for fumes to escape. Even normal filling can be polluting if gas stations in your area aren't required to use vapor-trapping devices. These capture the fumes given off by benzene and other hazardous chemicals in gasoline. (The devices are required in some areas.) Station owners don't like these devices because they cost hundreds of dollars each. And some consumers don't like them, either, because some of the more poorly designed ones make pumping gas less convenient. But the benefits to the earth are unmistakable. So, if you can choose between a station that uses vapor controls on its pumps and one that doesn't, choose the less polluting method.

Gas spilling out into the street may not be the worst of it. All vehicles have a built-in vapor recovery system. It feeds from the carburetor or fuel-injection system under the hood and into the gas tank. The system has little outlets near the gas tank fill neck. If you overfill the tank, you can put raw gas into an area that's only supposed to collect vapor. It can really mess up the system, particularly in a modern car that's computerized. Simply put, the system doesn't want to see liquids where there's supposed to be vapors.

DISPOSING OF HAZARDOUS SUBSTANCES

Your car is a repository of a variety of materials and fluids that can harm the environment: gasoline, gas additives, motor oil, transmission fluid, brake fluid, antifreeze, coolant, and battery acid. If you do some of your car's maintenance work yourself, the way you dispose of these fluids can have a severe environmental impact. Here is the proper way to dispose of common automobile products:

Batteries

Each year, about 80 million lead-acid car batteries are discarded, each containing about 18 pounds of lead and a gallon of sulfuric acid. About 80 percent of these are recycled, but 20 percent—16 million batteries—end up in landfills, where their lead can seep into groundwater. (Lead is a highly toxic substance that causes anemia, nerve damage, and paralysis. About two-thirds of the lead used in the United States is for lead-acid batteries.) In addition, deadly sulfuric acid leaks out in landfills and into groundwater.

Batteries can be recycled with ease. Recyclers open old batteries and drain out the sulfuric acid. It's either reprocessed or sent to a hazardous waste dump. Then the lead is removed and sent to a processor, who melts it into ingots. All told, it's a dirty, dangerous job, but it keeps some pretty toxic stuff out of our water supply.

Although service stations used to pay a few dollars for dead batteries, now some won't accept them at all. If they do, they usually charge you a few bucks for disposal. A few jurisdictions in the United States are beginning to require deposits on car batteries as a means of keeping them out of the local waste stream. But you can still usually find a service station that accepts batteries. Some hazardous waste facilities accept car batteries. If you change your own battery, get rid of the old one immediately. Don't let it sit around in your garage or basement. Improperly stored batteries can be a fire hazard.

If your local service station won't take it back, check with your city or county hazardous waste disposal office.

Motor Oil

Do-it-yourself oil-changers are not just a small group of backyard grease monkeys who read *Motor Trend* and *Hot Rod Monthly*. They're average working stiffs. In fact, at least half of all motorists change their own oil. That's maybe 50 million people.

The problem comes when it's time to dispose of the old oil. According to government statistics, 90 percent of oil-changers do one of two things with their used oil: they either dump it down the drain or throw it in the trash. Either method inevitably results in oil getting into the water supply. And both practices represent a shameless waste of two precious resources: oil and water.

The statistics, as pointed out earlier, are impressive: the oil equivalent of 35 Exxon *Valdez* tankers being dumped into our nation's rivers, lakes, and streams every year. The used motor oil you dump into the drain is far more deadly than the crude stuff that was leaked from the *Valdez*. Once burned in an engine, motor oil contains a deadly medley of poisons: arsenic, barium, beryllium, cadmium, copper, and on down through the alphabet to zinc. Exposure to used oil—for people as well as plants and animals—can have serious long-term consequences.

As we've seen all too often, it's not a pretty picture. Used oil is insoluble, persistent, and very slow to degrade. It sticks to everything, from beaches to birds. Moreover, an oily film on the water's surface blocks sunlight, which impairs photosynthetic processes and keeps oxygen from getting to fish and other aquatic life. Oil's toxic substances become concentrated in the tissues of plants and animals—some of which may be consumed by humans.

Aside from all of this, recycling oil makes good sense. For one thing, the oil can be re-refined—that is, turned back into oil suitable for use in cars or other machines (see box, page 81). The refining process isn't entirely efficient: One gallon of used motor oil yields 2.5 quarts of re-refined oil. But when you consider that it takes 42 gallons of crude oil to make those same 2.5 quarts of refined oil (there are some usable by-products left over), oil recycling seems like a pretty good bargain.

Whoever changes your oil, there are some important things you can do to help keep oil out of the waste stream.

Re-refined Oil

We've heard a lot lately about "closing the loop"—buying products made from recycled materials to help build markets for recycled materials and, ultimately, encourage even more recycling.

All of which would suggest that buying motor oil made from recycled material would be the right thing to do.

In the past, this wasn't necessarily the case. The first products made from recycled oil could be fatal to cars when poured into the crankcase. But new technologies are enabling processors to turn old oil into new products indistinguishable from those made from virgin petroleum.

Recycling oil does more than reduce pollution. It also cuts oil consumption. Motor oil consists primarily of "base lube," made from crude oil. The average content of base lube in crude oil is about 2 percent. So, to make a gallon of motor oil takes about 50 gallons of crude; the wasted 98 percent oil inevitably ends up being used for petroleum-based chemicals, fertilizers, or other products.

When motor oil is "spent"—when it's time to drain it and add fresh stuff—what gives out is not the base lube but the oil's additives. In addition, the oil becomes contaminated with traces of heavy metals from engine wear, and gasoline and other chemicals, all of which help to reduce its effectiveness.

The earliest methods of oil recycling involved filtering it through acid-activated diatomaceous earth. But the process created mountains of solid waste and the cleansed oil wasn't necessarily good for cars. But newer processes have resolved both problems.

Presently, motor oil with a purely re-refined content is available only through a few distributors; it is available in case quantities by mail order, usually for around $1.40 to $1.50 per quart. As more of us request it at service stations and auto supplies stores, it will undoubtedly become more accessible and less expensive.

How to Safely Dispose of Motor Oil. There are a few simple steps:

- Store the used motor oil in a clean plastic container with a secure lid, preferably a screw-on one. (You can get inexpensive, specially designed plastic containers at many auto parts stores.)
- Don't contaminate the oil with anything else. If it comes in contact with such things as window-washing fluids, air-conditioning solvents, gasoline, antifreeze, paint, other types of oil, or even water, it could spoil the oil and make it impossible to recycle.

• Take the plastic container of used oil to a recycling facility. In some cases, it may be as close as your nearest service station or repair shop, preferably one where you do business regularly. Because they change many cars' oil, they accumulate gallons and gallons of the stuff. Every week or so, a truck comes by and collects the used oil, which is sent to a re-refinery. But not all service stations and repair shops accept used oil, and some may charge a small collection fee. One reason is that the shop-owners have no idea what's in your oil, and whether you've let it come in contact with any contaminants.

Another possible oil drop-off site is a local auto parts store. Many now collect used oil. This is particularly true of stores that are part of national or regional chains. In fact, if your parts store doesn't accept used oil, you should take your business somewhere else.

About half of the states now have oil recycling programs. Many areas have city- or county-run oil collection sites where you can drop your container off. A few communities even have curbside pickup of used oil. Check with your city or county government or state environmental protection agency, or ask your local service station attendant. (If your community doesn't have a used-oil recycling program, consider starting one. For a free manual, "How to Set Up a Local Program to Recycle Used Oil," write to the U.S. Environmental Protection Agency, Office of Solid Waste, 401 M St. SW, Washington, DC 20460; 202-260-4610.)

• If you also change your car's oil filter, keep in mind that a used filter contains about a quart of oil. Experts recommend that you drain the used filter into your plastic container for at least 24 hours before disposing of the filter in your trash.

Remember that in most jurisdictions, improper oil disposal is illegal. In some states, you could be fined thousands of dollars and even face imprisonment.

Air-Conditioning Coolant

Automobile air conditioners are one of the leading sources of ozone-depleting chlorofluorocarbons (CFCs). CFCs are part of the coolant, the material that circulates within the air conditioning system to remove heat and generate cold air.

Any air conditioning or refrigeration coolant is supposed to be a closed system. The same coolant flows constantly through the

"Energy Conserving II" Oil

When you buy motor oil, there's something you should look for that can make your purchase a little greener. Look for the words "Energy Conserving II" on the container's label. This means the oil has special additives that help reduce friction and increase fuel economy. According to the American Petroleum Institute, these oils produce a fuel-economy improvement of 2 or 3 percent over a standard oil. That may not seem like much of an improvement, but if every car on U.S. roads got that much better gas mileage, experts believe we could save more than 400 million gallons of gas a year.

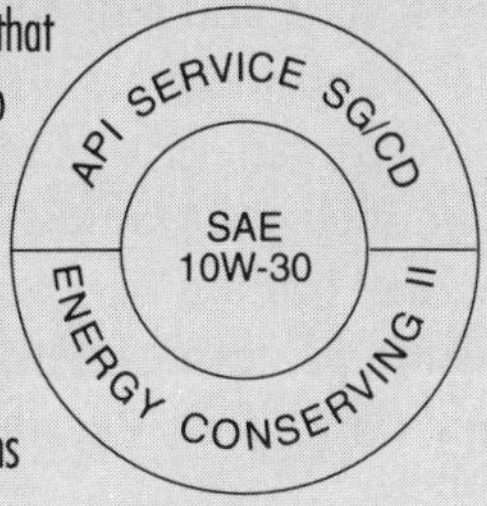

system. Just think about your refrigerator: How often do you need to add coolant, or otherwise repair the thing? If it's working correctly, probably not at all. Refrigerators sit quietly (or not so quietly) in the kitchen, doing their thing. With the exception of opening and closing it several times a day, it doesn't get a lot of abuse. And the temperature inside your house doesn't vary much, even from the dead of winter to the dog days of summer. So a refrigerator can work without failure for years and years.

Not so with a car's air conditioning. The unit is part of a noisy, vibrating engine, which moves along at a mile a minute. It's not only subject to the dead of winter, it reaches operating temperatures of hundreds of degrees. There's a lot of dirt and grime surrounding it. And every time you parallel park, tapping an adjacent car's bumper, it gets another jolt.

Given all this abuse, even the best system is prone to leaks. And what leaks is the CFC-based coolant, releasing ozone-depleting chemicals into the sky. So, car air conditioners need regular repair. How they're repaired makes a big difference to the environment.

Consider the act of having your car's air-conditioning system recharged—that is, having a mechanic add additional coolant to improve the air conditioner's effectiveness. Regular recharging indicates a leak in the system—after all, it's supposed to be a closed system—a leak that will not go away simply by adding more coolant. So this "repair" is really a sham, and an affront to the environment.

A growing number of shops are using recovery and recycling

equipment to capture and reprocess CFCs. The devices have come to be known as "vampires," probably because their preoccupation is to suck vital fluids. Simply put, vampires pump the refrigerant out of the compressor, clean and purify it for reuse, then store or transfer it while the air conditioner is being repaired. In coming years, state and federal laws will require all shops to use vampires.

The problem of CFCs in cars will eventually fade away as CFC-based coolants are banned internationally. Several auto makers are introducing air conditioning systems without CFCs. But none of the 130 million or so cars already on the road will be able to take advantage of this technology. So, we'll be stuck with dealing with cars' CFCs for a decade or more, until these cars are off the road.

Antifreeze/Coolant

Your radiator's antifreeze/coolant is most likely made from ethylene glycol. That's not horrible stuff: It's only moderately toxic and is thought to biodegrade in water relatively quickly. But that doesn't mean you or anyone else should drink it. More than a few family pets have died after drinking sweet-tasting puddles of antifreeze they find on driveways or in the street. The problem is that like used motor oil, antifreeze/coolant picks up a mixture of heavy metals and other toxic substances. So, simply flushing your radiator directly into the sewer system isn't a good idea.

Proper disposal is similar to that of oil: You pour used antifreeze/coolant into a sturdy container and bring it to a service station, or have it picked up at curbside, if there's such a program in your area. Second best is to dilute the antifreeze with as much water as possible and pour the mixture into a gravel pit or area with good drainage. (Make sure you are not violating any dumpign ordinances, however.) Don't do this if the fluid was used in a pre-1975 car. It may contain dangerous levels of lead. In that case, a service station or hazardous waste dump is the proper place for disposal.

In the end, proper disposal may take a bit of work. But whatever you might save in time and effort by irresponsible dumping is an illusion. In the long run, such practices will cost all of us dearly—in the money to clean up piles of tires and batteries and oil-contaminated waters; perhaps even in our own health, as we get sick from an ever-growing amount of toxic automobile debris.

THE GREEN COMMUTER

The idea of people sharing rides or taking turns driving isn't new—it's been going on ever since the first car-owner drove to work. But carpools and vanpools are becoming much more common. And with good reason: They offer a variety of benefits to participants and to the environment.

But there's another reason for their growing popularity. The Clean Air Act of 1990 mandates that the biggest employers in the most polluted cities must take actions to help adopt "transportation control measures" to reduce the number of "vehicle-miles" traveled by its employees. And states are getting into the act, too. Washington state has required that major employers in the eight most populous counties cut their number of solo commuters by 35 percent by 1999. The companies must draw up a plan or face civil penalties. Employers can do anything that works, from offering mass-transit subsidies to charging for parking.

"Transportation control measures" include a variety of things, including setting up high-occupancy vehicle (HOV) lanes, restricting vehicle use in congested areas during peak periods, and even limiting driving altogether if necessary. There are about 400 miles of HOV lanes throughout the United States, with another 500 or so miles planned. A study at the Texas Transportation Institute found that a HOV lane, which can be reversed with the flow of rush-hour traffic, can save 6,200 gallons of gasoline and 13,420 pounds of carbon monoxide a day compared with adding a new lane in each direction. Seattle, which is embarking on an ambitious plan to build nearly 300 miles of HOV lanes, is combining them with special on-and-off ramps, preferential parking, and other incentives.

Obviously, simply building HOV lanes isn't enough. For them to work, commuters will have to abandon their solo driving habits and team up with other commuters to form carpools and vanpools.

CARPOOLING

It only takes two people to form a carpool—just a couple of neighbors or co-workers who have recognized that there are a host of benefits to be gained from pooling one's resources, literally. Carpools save money, pollution, time, and stress, among other things. They can even make you more productive at work.

According to the Association for Commuter Transportation (ACT), there are two basic types of carpool arrangements:

• The poolers use one car owned by one driver. The driver calculates his or her operating costs for the daily commute, then divides the cost by the number of riders to determine how much each rider should contribute. The riders and driver agree to a periodic (weekly, monthly, daily) payment plan.

• Or, poolers can rotate car use and driving, so that each person's vehicle and time is shared equally. No money is exchanged in this arrangement.

Setting up carpools is pretty easy. A growing number of communities have set up formal networks to facilitate carpooling (also known as ridesharing). The phone numbers are sometimes posted along highways: 555-RIDE, or 555-POOL, for example. If there's no formal network in your community, try posting a notice at a local community bulletin board—at a church, community center, supermarket, or other gathering place. Or circulate a memo at your company or at nearby companies, seeking riders or drivers.

See if your employer is willing to get involved. Companies are recognizing that many carpoolers come to work more refreshed and less stressed than those who have to fight traffic every day. A lot of employers will help to organize carpools by providing forms, a bulletin board, or other means.

Here are a couple of other ideas:

• If your company has an employee parking lot, suggest to someone in charge that special close-in spots be reserved for carpools. Some companies charge employees for parking, but give discounts or charge nothing to carpoolers.

• If your employer will allow, post a large map of the area around your company on a bulletin board. Distribute flagged pins to each employee in carpooling and request that they write their name and phone number on the flag. Then have them stick the pins on the map where they live. That will provide a graphic illustration of who lives near whom, allowing neighbors to set up carpools. You could go one step further and organize a list of employees by neighborhood, then distribute it with phone numbers to would-be poolers.

Keeping the Pressure Up

Who says the little things don't count? Sometimes, as we view the enormity of the environmental problems of our world, we forget about how some of the little things can make a big difference.

Consider automobile tire pressure, for example. If you are a typical American driver, you probably are driving around with underinflated tires. That may not seem like a major environmental blunder, but in fact it is making a significant contribution to global warming. The basic assumption has to do with the fact that each gallon of gas your car uses contributes about 19 pounds of carbon dioxide (CO_2) into the air.

According to the Environmental Protection Agency, at least 65 million cars on American roads have underinflated tires. That low air pressure increases a car's rolling resistance, decreasing gas mileage by up to 5 percent. If you assume that these 65 million cars get an average of 18 miles per gallon (that's the EPA's estimate), that each drives 10,000 miles per year, and that 5 percent of the gas is wasted due to underinflated tires, that comes to *17 million tons* of CO_2 released into the air simply because of low tire pressure.

Once commuters find a good match—one or more people whose schedules and locations are compatible—the next step is to talk on the phone, and sometimes meet briefly in person, to discuss the details of their arrangement. It's wise to set a trial period of a week or two, during which new poolers can become acquainted and determine whether their personalities and preferences will mesh satisfactorily.

There are many benefits to carpooling. Aside from conserving fuel and cutting pollution, carpools reduce traffic congestion and subject roads to less wear-and-tear. Even better, carpools allow some riders to drive their cars less, reducing repair bills and extending their vehicles' lives. And, of course, pooling allows riders to read, talk, sleep, or simply think without having to deal with congested traffic. That can make you a happier, healthier, more productive person.

VANPOOLING

Vans hold more people than cars—as many as seven to ten passengers—so vanpooling is a slightly more structured and organized form of carpooling. Like carpooling, there are several options: a vanpool owned and operated by a commuter who wants to have riders share his or her cost of operating the van, an employer-sponsored vanpool, or a third-party vanpool.

The first type is simple enough. If you owned a van, you could arrange to drive others to work—in effect, creating a carpool in which you drove every day. By charging a nominal fee to passengers, you would defray your expenses and possibly pay for the van itself. Of course, you would be responsible for the van, including all gas, maintenance, and repairs, as well as insurance and other costs.

The employer-sponsored vanpool uses vans owned (or leased) by an employer and offered to employees for commuting as an extra benefit. Companies such as 3M, Bechtel, Hughes Aircraft, GEICO, and many others operate fleets of vans to help employees get to work without driving their cars. In most instances, employees pay their employers a set fee, depending on the length of the commute. The employer maintains and insures the vans.

In some communities, private entrepreneurs set up vanpooling operations, available to employees or companies on a month-to-month basis. The van is kept by the employees for commuting purposes, but the private company maintains and insures the van.

Because vanpools require special vehicles—vans—setting them up is a bit more involved than forming a simple carpool. First, there's the cost of obtaining, maintaining, and insuring the van. Then there's the matter of who will drive, who will collect the fares, how much the fares will be, and so on. Local or regional carpooling organizations may be able to provide some assistance.

Local Sources of Information

• To obtain ridesharing information for your local area, send a stamped, self-addressed envelope to ACT, 808 17th St. NW, Ste. 200, Washington, DC 20006. They'll send a list of ridesharing contacts. Also available from ACT is a national commuter resource directory, containing nationwide ridesharing contacts, as well as federal resources. The guide is available for $35 from ACT.

• For vanpooling information, contact VPSI (P.O. Box 159, Detroit, MI 48288; 800-223-8774) to find a local office. VPSI sets up vanpools nationwide by leasing vans and providing maintenance, repairs, financing, insurance, and administrative services.

CLOTHING & PERSONAL CARE PRODUCTS

While the cosmetics, clothing, and personal care products industries have long been concerned with the ecologies of our bodies, they have only recently begun considering that of our planet. The past two decades have seen a remarkable growth in products intended to make us feel good while they make us look good, too. (Indeed, a good case can be made that the two go hand in hand.) But whereas yesterday's products took on a look-good-whatever-it-takes attitude, many of today's products place a premium on what effects they might have on our short- and long-term health.

What about the environment? There's some progress there, too, but not much. One of the biggest problems is packaging—the creators of cosmetics and health products are masters of it. From the glorious shapes of bottles to the convenience of pocket- and purse-size objects, the purveyors of everything from shampoos to shaving cream to suntan oil have tried to accommodate our every need. And in the process, they have contributed significantly to the mountain of waste we create, and to some of the harmful matter this trash releases into our air, water, and soil.

Another issue for some Green Consumers involves the process involved in getting many of these products onto our shelves. Throughout the world, there has been growing concern about the use of animals in the safety testing of all types of products. Their use in cosmetics has been particularly controversial because these products are considered nonessential. To address this increased concern, a growing number of companies have produced what have come to be called "cruelty-free" products—those that do not involve animal testing or that use animals in a responsible manner.

The good news is that, on all fronts, Green Consumers have an increasing number of choices available of quality products that don't harm the earth—or any of its creatures.

THE FABRICS OF OUR LIVES

Which type of fabric is the greenest—the one that causes the least amount of harm to the environment? If you're like most people, your initial answer might be "anything that's natural and not synthetic." And that's a reasonable guess. After all, cotton and wool are made from plants and animals, while most synthetics come from petroleum-based chemicals.

But you might also be surprised to find that "natural" fabrics are far from environmentally benign. In fact, the more you understand about how the materials used to cover our beds, bodies, windows, and floors come to market, the more you learn how much even natural materials can be hazardous to the planet.

Let's start with cotton. At first blush, it would appear to be the most benign of materials because it comes from plants, not animals.

But growing those cotton plants is a deadly business. "Fifty percent of all pesticides used in this country are applied to cotton," says Stanley Rhodes, president of Scientific Certification Company, which is conducting life-cycle analyses of cotton products. "It has a tremendous impact on the environment." Moreover, some of the pesticides, herbicides, and fertilizers used to grow cotton are derived from petrochemicals. All told, conventionally grown cotton can be as petroleum-intensive as nylon, according to research conducted by Dr. Paul Demko of the Good Housekeeping Institute.

Petroleum isn't the only problem. Nearly 18 million pounds of insecticides are dumped on American cotton fields each year, according to the U.S. Department of Agriculture. Those pesticides include some of the most lethal ingredients around, including aldicarb and methyl parathion, both suspected carcinogens. Not only are they a risk to the environment by poisoning wildlife and seeping into groundwater, they also can affect farm workers. "We really are poisoning our people," says Terry Gips, president of the International Alliance for Sustainable Agriculture.

Herbicides are another threat, applied both to control weeds and to defoliate the cotton before harvesting. A majority of cotton farmers do use nonchemical pest-management methods, such as hiring workers to physically remove pest insects from plants, but usually in addition to the use of chemical pesticides.

That's only the beginning. Once cotton is harvested, it undergoes another chemical feast to turn it into fiber. For example, cotton lint typically is bleached in a hydrogen peroxide solution, which gives the material a uniform color and washes away the waxy substances naturally found on the fiber. While hydrogen peroxide isn't much of a threat—it readily breaks down into water and oxygen—the process for manufacturing the stuff is quite energy intensive.

Conventional methods for processing cotton involve another twenty or so steps, some of which employ harsh chemicals such as chlorine bleach. The bleach can react with cotton to form dioxins, which are thought to be cancer-causing. Formaldehyde, a suspected carcinogen, is used to give some cotton its permanent press qualities. Throughout the fabric's life, consumers are subjected to slow, continuous release of this chemical, which can be irritating to eyes, lungs, and skin.

There's more. Preshrinking cotton can use liquid ammonia, which, like peroxide, is energy intensive to make. Mordents—chemicals used to make dyes adhere better—can also be toxic.

"Greener" Cotton

Fortunately, there is hope for those who prefer cotton. Some farmers, wholesalers, and retailers are beginning to respond to a demand for "green"—or at least "greener"—cotton products made using methods that are less energy- and chemical-intensive.

Seventh Generation (Colchester, VT 05446; 800-456-1177), the mail-order company, is among the bigger boosters of products made from what it calls "green" cotton. The company's bath towels and clothing are made from material free of bleaching or other chemicals used in the advanced processing steps. Moreover, the company recently has located organic growers, which will enable Seventh Generation to eventually sell cotton grown without any pesticides or chemical fertilizers.

Chemical-free cotton may take some getting used to. For one thing, because it is not bleached, some of the natural waxy substance of cotton remains in the material, at least for a while. Says Seventh Generation's Martin Wolf, compared with conventional towels, "Our towels are not as absorbent when first purchased, but they are fine after a couple of washings."

There's a price to be paid for this new, greener technology. Seventh Generation's "green" cotton products are a bit more expensive than their conventional counterparts. In its upcoming fall catalog, a set of two sheets and pillowcases for a standard twin-sized bed will go for $42, about $5 more than a comparable brand-name department store version. (That's a big improvement. In its spring catalog, those same items sell for $123.) The higher prices, say the experts, has mostly to do with the higher costs associated with smaller growers of cotton. As organic cotton is grown on a larger scale, prices will theoretically drop.

There may be a happy medium found in "Simply Cotton," a new line of cotton products being introduced by J. P. Stevens, one of the nation's largest manufacturers of bed and bath products (1185 Avenue of the Americas, New York, NY 10036; 212-930-2000). The new line, packaged in recycled cardboard, includes unbleached and undyed sheets, pillowcases, bath towels, and bathroom rugs. The company is issuing the new line in response to "tremendous consumer demand for this kind of product," according to public relations director Rose Gerace. Already, she says, initial sales have greatly exceeded expectations. One reason may be that the suggested retail price is comparable to Stevens' designer sheets, despite higher production costs.

Wool

What about wool? It, too, comes from renewable resources—sheep—and requires far less processing than cotton. But wool is still less than perfectly green.

Primarily, there is the animal-protection issue. "If you are wearing wool, you are supporting a system that has a long tradition of wiping out magnificent predators," says Robin Duxbury, executive director of the Rocky Mountain Humane Society. Duxbury and other environmentalists are fighting sheep ranchers in Colorado and other western

Hemp and Glory

Is marijuana a source for environmentally sound fabrics? Before you scoff at the notion of smokable sheets and shirts, it's worth taking a closer look at this controversial—and often underestimated—plant.

According to *The Emperor Wears No Clothes*, by Jack Herer, 80% of all textiles and fabrics for clothes, linens, rugs, sheets, and towels were once made principally from cannabis fibers, or hemp—the marijuana plant—until the 1820s in the U.S. and the early 20th century in most of the rest of the world. As Popular Mechanics put it in 1938, "Hemp is the standard fiber of the world."

Apparently, hemp is an easy crop to grow. It has a short growing season and can be cultivated practically anywhere. The marijuana lobby, once content to decriminalize the weed, has launched a new campaign to show that hemp has a myriad of valuable—and ecological—uses. Among other things, they say, it can be made into paper (the maps Columbus used to "discover" America were said to be printed on it) and is the basis for fuels for heating and transportation that could ease global warming.

Obviously, there are other issues here, including complex legal ones, but any thorough discussion of the environmental impact of fabrics would be incomplete without at least considering the potential of hemp as a good, green alternative.

For more information, contact the National Organization for the Reform of Marijuana Laws, 1636 R St. NW, Washington, DC 20009; 202-483-5500.

states—the source for nearly all U.S. wool—who are trying to get the puma, or mountain lion, reclassified as a "varmint" because it has made a significant rebound from near eradication during the last few decades, when wildlife authorities put the species under protection. Ranchers, say environmentalists, prefer to poison or shoot puma, coyotes, and other sheep predators rather than using less destructive methods, such as guard dogs.

The sheep industry, of course, sees things differently. "The industry has never backed a policy based on eradication, merely predator control," says Janice Grauberger of the American Sheep Industry. "Guard dogs work well as a deterrent, but they're not perfect. Our members are just trying to protect their means to make a living. Sheep are an earth-friendly product that provide both food and fiber. They graze on broadleaf plants that are not consumed by any other livestock. They are using resources that otherwise go to waste."

Synthetics

Of course, there are dozens of synthetics, materials typically made from cellulose fibers or petrochemical-based resins. Surprisingly, the synthetics industry can make a good case why their fabrics compare favorably, environmentally at least, with cotton and wool.

For one thing, synthetics use less than 1 percent of all petroleum used in the U.S., according to Tom Lukiash, environmental affairs manager at E. I. DuPont de Nemours, Inc., the world's largest synthetic fabric maker. Moreover, synthetics are recyclable—theoretially, at least—and about 10 percent of them are recycled, says Lukiash. But much as with plastics, what hampers recycling efforts is the lack of an infrastructure. Simply put, there are precious few recyclers able and willing to accept synthetics. The reason: There are only a handful of recycling plants up and running. Nearly all synthetics recycling comes from big industrial users, who are able to ship scraps or other unused fabric back to manufacturers, who then reprocess the material.

Another benefit of synthetics is their durability. "Nylon can last as much as 25 times longer than cotton," says the Good Housekeeping Institute's Demko. "That makes it less resource intensive." Moreover, tests have shown synthetic fibers to be less susceptible to rot and mildew.

Clearly, manufacturing synthetics can be highly polluting. In 1987, for example, DuPont used more than 20 billion pounds of raw materials to produce just 3 billion pounds of manufactured fiber. The process yielded nearly 14 pounds of hazardous waste. Despite improvements—last year, the equivalent production created less than half the amount of hazardous material, and new processes promise further reductions—synthetics' air pollution is thought to be much higher than that of either cotton or wool. Lukiash disagrees. Because the chemical industry is more highly regulated than is agriculture, he says, pollution controls are tighter for synthetics. "It's not that cotton producers don't generate pollution, it's that ours is known."

What's Best?

We're certainly not suggesting that petroleum-based synthetics are the green answer. There are other considerations—comfort, appearance, durability, and the suitability of the fabric for the application—

that may make cotton or wool preferable. And then there are blends of natural and synthetic fibers, which may offer the best of both worlds. What's important is to understand that all materials have their environmental impact, and that traditional arguments—that "natural" materials like cotton and wool are always environmentally preferable to petroleum-based synthetics—are simplistic, to say the least.

Clearly, there is much more to learn. And as life-cycle studies are conducted for each of these materials, we'll eventually learn which fabrics—and which companies making products from these fabrics—offer the better solutions for the environment.

Green Cotton Manufacturers

One of the greatest boons to the green cotton market is Fox Fibre, a means of naturally coloring cotton. The invention of entomologist and yarn spinner Sally Fox, the idea came to her from some ancient textiles she saw in a museum that were naturally colored. After considerable trial and error, she was able to grow cotton in two varieties: brown and green; there's also a natural, cream-colored version. Even better, the colored cotton is more resistant to pests, so it requires fewer pesticides than conventional cotton, and is much easier to grow organically. Bleach is left out of the process, another environmental plus.

A growing number of cotton products, including many listed below, are made of Fox Fibre.

• **EarthCare, Inc.** (P.O. Box 326, Hickory, NC 28603; 704-327-3633) specializes in what it calls "environmentally safe" socks. Unlike conventional socks, which use nylon or spandex to maintain the stretch, EarthCare uses natural rubber thread wrapped with cotton yarn. The company claims to be the only manufactureer using cotton yarn amd adapted machinery in order to successfully hold the toe seam, as opposed to the nylon normally used to reinforce the seam.

• **Earthlings** (205 S. Lomita Ave., Ojai, CA 93023; 805-646-7770) uses Fox Fibre naturally colored cotton in shades of brown and green for their line of shirts, paints, dresses, hats and jumpers. Earthlings uses organically grown, formaldehyde-free, unbleached cotton, as well as rainforest-grown Tagua Nut buttons. The line is available at specialty stores and through the Pure Podunk catalog (see next page).

• **Ecosport** (28 S. James St., Hackensack, NJ 07606; 800-486-4326) was founded by two environmentally concerned parents in their effort to create a sportswear line — everything from T-shirts and sweatshirts to baby's crib sheets and blankets — from all-natural fibers, manufactured without bleaches, dyes, and harmful chemicals.

• **Esprit** (900 Minnesota St., San Francisco, CA 94107; 415-648-6900) offers its Ecollection, a clothing line that incorporates green manufacturing products and processes, from buttons (carved from nuts, handpainted wood, or reconstituted glass) to the cloth itself (some organically grown, some vegetable dyed). Knitted fabrics are mechanically—as opposed to chemically—pre-shrunk to eliminate resins and formaldehyde. Ecollection products, which include pants, shorts, jackets, and T-shirts in earth tones, are now available in Espirit stores worldwide.

• **Levi Strauss & Co.** (P.O. Box 7215, San Francisco, CA 94120; 1-800-USA-LEVI) has created a Naturals line of denim products for men and women, fabricated from naturally colored Fox Fibre. Products in the Naturals line include "550" relaxed-fit jeans, slouch jackets, and knit shirts for men, and "512" slim-fit jeans for women. Both lines come in the color coyote brown. Call the toll-free number to locate the Levi's Naturals retailer in your area.

• **O Wear** (P.O. Box 77699, Greensboro, NC 27417-7699; 919-547-7886) is a line of casual knit clothing made only from organic cotton. O Wear offers textured-knit fabrics in a variety of colors, with styles designed for a relaxed, body-conscious fit. O Wear supports organic farming through its contracts with conventional farmers to make the transition to growing organic cotton. After two years of organic farming, the crops can then be certified 100-percent organic.

• **Patagonia** (Lost Arrow Corp., P.O. Box 150, Ventura, CA 93002; 805-643-8616) acknowledges that every piece of clothing made has a negative impact on the environment. This mail-order sporting-goods company has instituted an "Environmental Review Process" to examine the materials and methods used in the production of its merchandise. Patagonia's founder helped to start a corporate conservancy group, the Outdoor Industry Conservation Alliance, which develops programs and donates money for environmental causes. Above all, Patagonia encourages consumers to buy only the

Just Redo It

Nike has announced a sneaker made of recycled materials. The country's largest athletic shoe maker said it will be offering a new version of its Air Escape Low shoes with soles made from about 20 percent recycled materials. Plans call for the company to have recycled soles on all seven styles it makes for outdoor sports.

Meanwhile, an Oregon entrepreneur seems to have gotten her foot in the door with a good idea: DejaShoes, footwear made from recycled materials. The soles come from 100-percent reclaimed tire rubber. The inside and outside fabric are made from post-commercial polypropylene—specifically, scrap from hospital gowns and diapers. And the padding is made from 100-percent reclaimed foam rubber. Like Henry Ford's Model T, the $49.95 shoe is available in just one style and color: a tan, lace-up women's walking shoe. Fortunately, it's available in several sizes: 5 to 11, including half sizes and narrow, medium, and wide widths. DejaShoes are available from only a few retailers, but you can buy them direct by calling 503-624-7443.

items that they need and to shop for long-lasting, quality merchandise to provide maximum wear and minimum waste.

• **Pure Podunk** (Old Schoolhouse Ctr., P.O. Box 194, Sharon, VT 05065; 1-800-PPODUNK) is a mail-order firm that caters to customers with severe chemical sensitivities. It also encourages consumers to contribute to its "Financial Aid Fund" for chemically sensitive people who need pure products but lack the funds to buy them. Pure Podunk's catalog contains products from several companies, including toys, furniture, bedding and clothing.

• **Robin Kay** (201 Dufferin St., 3rd Fl., Toronto, Ontario M6K 1Y9; 416-532-7899) knits sweaters, skirts, leggings, and dresses in a wide variety of stitches and styles. Its collection also includes professional and evening wear. The company's operations include a knitting mill, cut-and-sew operation, and eight retail outlets. The company uses unbleached, undyed cotton it calls Supernatural.

• **Utica** (1185 Ave. of the Americas, New York, NY 10036; 212-930-2617) Simply Cotton, a line of all-cotton sheets, bedroom accessories, bath towels, and rugs free of chemicals, additives, bleaches, and dyes. The products come in natural tones and are packaged in a biodegradable paper box made of recycled material.

• **Wearable Integrity** (1725 Berkeley, Santa Monica, CA 90404; 310-449-8606) specializes in casual knitted sportswear for women, as well as a version for children. In all products, fabric processing is kept to a minimum, including the elimination of unnecessary chemical treatments. The clothes are natural colored with Fox Fibre colors.

• **Wearables Marketing** (401 Ocean Front Walk, Venice, CA 90291; 310-399-5608) offers a line of "PurelyGrown" socks. The firm contracts with American farmers to grow cotton under strict organic regulations. The socks are manufactured without formaldehyde, bleaches, dyes, or heavy metals, and are colored with Fox Fibre in three colored: leaf green, earth brown, and harvest cream.

THE DIAPER DILEMMA

You've come a long way, babies.

Not very long ago, disposable diapers were seen as the bane of the environment. It was easy to see why: American families each year toss out more than 18 billion diapers, containing an estimated 2.8 million tons of excrement and urine, into America's dwindling landfills. That amounts to almost 44 million diapers a day, or more than 500 every *second.* The typical baby soils some 7,500 disposable diapers before graduating to the toilet. Like almost everything else that ends up in landfills, the diapers' impact on public health and the environment is unknown.

Some facts about disposables are well established. They consume some 67,500 tons of polyethylene resin each year, which is used in the waterproof plastic liners of most commercial brands. When disposed of, that plastic takes hundreds of years to break down. (Or so researchers say. No one really knows; these diapers have only been around for about a quarter-century.) The figures do not include the other synthetic components of most disposable diapers, including the diapers' elastic, tape tabs, adhesive, and frontal tape. Disposable diapers represent just under 2 percent of all household solid wastes—the largest single product in the waste stream after newspapers (6.8 percent) and beverage containers (5.5 percent). All told, American babies produce an estimated 3.6 million tons of used diapers a year.

There's more. Even though parents are instructed to flush the soiled inner lining of the diaper down the toilet, one study found that

Other Disposables

Here are some other matters of concern to Green Consumers regarding disposable personal care products:

• **Razors:** Americans go through about two billion disposable razors each year—the kind that include a disposable plastic handle and blade in one. Clearly, these are a waste of resources (and money). These aren't recyclable; there's no known way to separate the metal from the plastic for reuse. Because of the ready availability of reusable handles (even if they have disposable blades), the disposable variety should be avoided.

• **Tampon applicators:** They have come to be called "New Jersey seashells": those plastic applicator tubes that come with some brands of tampons. When disposed of these seemingly harmless devices often end up in rivers and streams, and, eventually, oceans. There they are mistaken for food by birds and fish. When they are consumed, they often result in these creatures choking to death.

fewer than 5 percent do so. An infant's feces and urine can contain any of over 100 viruses, including live polio and hepatitis from vaccine residues. These viruses are potentially hazardous to sanitation workers and to others, both through groundwater or when carried by flies.

So, disposables are to be avoided by any environmentally conscious consumer, right?

Maybe not. The answer lies in the trade-offs between disposables and reusable cloth diapers. It turns out that each type creates its own environmental stress. Says Dr. Allen Hershkowitz, senior staff scientist with the Natural Resources Defense Council: "This is a complicated issue, and there's really no pat answer."

Here are some of the trade-offs:

• **Plastic diapers,** as we've already pointed out, require more materials to manufacture and generates far more solid waste in landfills. In addition, the paper fibers used in disposables are bleached using a process that generates dioxin contaminants that pollute water. (However, there is no known threat of dioxin poisoning of babies who wear the diapers.)

• **Cloth diapers,** for their part, have their own environmental costs. Studies have shown that the production of reusable diapers actually consumes more energy and water and generates more water pollution than the production of disposables. And then there is the water, energy, detergents, and disinfectants used in rinsing out, washing, and drying cloth diapers. If you use a diaper delivery service (about 10 percent of Americans who use cloth diapers do), you must also consider the pollution and energy use of the delivery trucks.

In the end, when you compare the resources used and pollution generated in the manufacturing, use, and disposal of both types of diapers, there is no overwhelming evidence that either type is better.

Of course, there is also the issue of which type of diaper is better for babies. According to Hershkowitz, "the various chemical, enzymatic, microbial, and physical conditions that diapers produce aren't favorable for human skin." The result is often diaper rash, which can make a baby miserable. Avoiding diaper rash requires keeping the baby's skin as dry as possible (which is why some parents use baby powder). Hershkowitz points to research that concluded that disposable diapers keep skin drier, especially if they contain an absorbent gelling material, a substance found in about 90 percent of all disposables.

Does that mean disposables are better? Maybe, but maybe not. Counsels Hershkowitz: "I think that it really has to be a personal choice—one you should make after considering the environmental and health factors and applying them to your own life. But be sensitive to other ecological problems. For example, if you are living in the Southwest, where water is a big problem, you might consider using disposable diapers. On the other hand, if you live in the Northeast, where landfills are scarce, perhaps a cloth diaper makes better sense, assuming it works for you baby's skin type."

Comparing Costs. Diapers are unique among disposable plastic products: They are one of the few products in which the disposable product, besides contributing to a host of environmental problems, costs more to use than a reusable product—in this case cotton diapers from a delivery service. On average, according to a study conducted by Energy Answers Corporation, a waste management consulting firm, cotton diapers cost about 15 cents each from a delivery service, compared with 22 cents for each disposable diaper. That seven-cent difference may not seem like much, but when you consider that babies require between 6,000 and 10,000 diapers over a three-year period, the difference amounts to between $420 and $700. With cloth diapers, human waste is properly channeled into the sewage system and the

The Bottom Line

The following data, provided by Environmental Action, are based on a child requiring eight changes a day over a 30-month diapering period—a total of 7,300 diapers:

• **Disposable diapers** can be used once each. At an average cost of 21 cents per diaper, the total for 30 months comes to $1,533.

• **Diaper services** permit a diaper to be used about 150 times each. At a cost of $7.50 per week, the total for 30 months (130 weeks) comes to $975.

• **Washing cloth diapers at home** permits a diaper to be used about 90 times (home washing typically involves more chlorine bleach than diaper services use, thereby decreasing diaper life). That requires about seven dozen diapers at an average cost of $9.23 per dozen, which comes to $64.61. Home laundering costs, calculated at 3 cents each (including detergent, water, electricity, and the depreciated cost of the appliances), come to $219, for a total cost of $283.61.

fabric can be reused up to 200 times before the fabric is recycled. Almost all worn-out diaper service diapers are recycled into rags for industrial use.

One barrier to using cloth diapers is day care centers. Many require parents to provide a supply of disposable diapers for their children. But a growing number of day care centers are rediscovering diaper services, too, and are including the cost of the service in parents' bills. **Kindercare**, a national chain of day-care centers, allows parents to provide cloth diapers if they submit a doctor's note stating that their child requires cloth diapers.

There have been some innovations that make the traditional washable cloth diaper even more convenient, comfortable, and durable. Contact these companies for their latest catalogs:

• **Baby Bunz & Co.** (P.O. Box 1717, Sebastopol, CA 95473; 707-829-5347) offers a wide range of "natural diapering" products, including Nikkys—breathable, natural fiber diapers that come in either cotton or wool felt.

• **Biobottoms** (Box 6009, Petaluma, CA 94953; 800-766-1254) makes Velcro-fastened, fitted wool diaper covers that help keep babies extra dry. The breathable covers can go through about five diaper changes before needing washing.

• **Bumkins** (1945 E. Watkins St., Phoenix, AZ 85034; 602-254-2626) makes a one-piece diaper that will last through 200 washings. It features a waterproof nylon cover and absorbent thick cotton padding.

Another alternative to buying cloth diapers is to use a diaper service. They are listed in the Yellow Pages under "Diaper Services." If you cannot locate one easily, contact the **National Association of Diaper Services**, 2017 Walnut St., Philadelphia, PA 19103; 215-569-3650.

EARTH-FRIENDLY COSMETICS

Ozone Safe . . . Environmentally Friendly . . . All Natural. Cosmetics and personal-care products makers are some of the heaviest users—and abusers—of green claims. What do those labels mean, and how do you know they're telling the truth?

The short answer is that you can't. But with a little perception and some common sense, you can figure out which, if any, products may be better for the environment—and for your health.

It's not always easy to determine the environmental impact of a particular product, but as with other packaged goods—from laundry detergent to microwave soup—there are some things to look for.

The most obvious is packaging. Cosmetics are heavily overpackaged. Manufacturers apparently believe consumers—especially women—like them that way. Look at any product in which there are separate versions for women and men. The women's version will inevitably have at least one more layer of packaging than the men's. (Mennen Speed Stick and Lady Speed Stick deodorant is one example: The women's version not only is overpackaged, ounce for ounce it's more expensive.)

There are alternatives. Tom's of Maine and products from The Body Shop are two examples. Tom's uses recycled paper in its cardboard boxes for toothpaste and other items; the cardboard contains at least 50 percent post-consumer waste. For their plastic bottles, both Tom's and The Body Shop use high-density polyethylene (identified by the number "2" on the bottom), one of the more commonly recycled plastics. And both companies have refill programs: Tom's of Maine sells refills for its stick and roll-on deodorants. The Body Shop, which has more than 80 U.S. stores, features a refill

New, Improved Aerosols

The environmental problems associated with aerosol cans did not end in the 1970s, when the federal government banned the use of ozone-destroying chlorofluorocarbons. The CFC ban led manufacturers of deodorants, hair sprays, and other aerosol products to substitute CFCs with chemicals that do not damage the ozone layer. Unfortunately, many of the substitutes contribute to the hydrocarbons (mostly propane, butane, and isobutane, which are not active ingredients but help force the product out of the can) that help create smog in many urban areas. Like hydrocarbons from a car's tailpipe, they combine with nitrogen oxide emissions and sunlight to form smog.

But new, improved aerosols are coming. In fact, two cutting-edge aerosol products have already be introduced. One is the Atmos "gas-free aerosol delivery system," from Exxel Container in Somerset, NJ. It resembles a pump-spray bottle, but when the top is pressed, it emits a constant mist much like an aerosol. Another advantage of the Atmos over aerosols is that it is at least partly recycled: While the inner "power assembly" is disposable, the outer container is recyclable in some areas.

The Atmos is already showing up on shelves. Among companies using the bottle are Chanel, Estee Lauder, and Sergeant's Pet Products, which recently introduced a flea and tick spray using the dispenser.

Equally impressive is the Biomat Airspray bottle, which requires the user to "pump" the applicator to add compressed air before spraying. Used in Europe for six years, it has just been introduced in the U.S.

Biomat's plastic bottle can be filled and refilled with almost any liquid, including cleaners, hair sprays, deodorants, even paint. According to Wolf Ebel, president of Biomatik USA Corp., the bottles have been tested up to 10,000 strokes. That means that a single Biomat could replace the equivalent of at least 500 regular aerosol cans, he says.

Biomatik expects to announce one or more major product introductions in coming months using the Biomat, but in the meantime, the dispensers themselves are available for home use. So far, they are sold mostly in health food stores, but they can be purchased by mail for $4.99 each, plus $3 shipping and handling from Biomatik, P.O. Box 2119, Boulder, CO 80306; 800-950-MIST.

Both of these product introductions will likely be overshadowed in coming weeks, when megalith Procter & Gamble introduces Vidal Sassoon Airspray hair spray in the U.S. The same product, already sold by P&G in Germany under the name Shamtu, is similar to Biomat, requiring a few "pumps" each time before using.

In any case, it appears that the aerosol may be making a turnaround, from a polluting, unrefillable, and nonrecyclable container to one that potentially meets all three of those criteria. That's progress.

bar where customers can bring in clean containers (including their own bottles) and fill them at a discount.

At beauty salons that sell products made by Aveda Corp., customers can return containers for recycling. Aveda also uses glass for many of its containers, except for its bath care products, which use plastic.

But beware: Green claims may be exaggerated, especially in products sold only in salons. According to Don Davis, editor of *Drug & Cosmetic Industry* and *Cosmetic Insiders' Report*, hair and skin products may contain as little as 25 percent recycled plastic, even though the marketers simply call their packages "recycled." (There are no legal requirements for using this term.) He adds that $350 million worth of hair and skin products were sold directly through salons last year. That's only part of the picture: Out of the $18.6 billion-a-year personal care industry, natural products accounted for $351 million in 1990.

Chemicals and Pollutants

That's just the packaging side of the question. There are environmental impacts when cosmetics and other products are manufactured, but those effects are harder to discern. According to EPA's data, makers of perfumes, cosmetics, and other toilet preparations released 2,465,800 pounds of chemicals into the environment in 1991. That by no means puts these firms in the same league with bigger polluters. By comparison, the petroleum industry, which ranked seventh among the nation's 25 most polluting industries, released more than 117 million pounds of toxic chemicals.

Still, there are compounds in cosmetics which aren't good for the earth. Methylene chloride, an ozone depleter and a probable human carcinogen, was banned from hair spray in 1989. But other pollutants remain. California, with its poor air quality and aggressive approach, has taken a strong position on aerosol hair-styling products, deodorants, and perfumes, which contain hydrocarbons and solvents that contribute to smog. Hair sprays in California account for 9 percent of non-automotive emissions of volatile organic compounds and at 46 tons a day, are the biggest source of emissions from consumer products, excluding spray paints.

It is difficult to trace what happens when shampoo, soap, and assorted oils, creams, and ointments are washed down the drain. If they go to a municipal wastewater treatment facility, the effect is probably not large. But there is an impact; after all, that's why biodegradable,

Head and Shoulders Above

Several companies are creating two-in-one hair-care products, combining shampoos with conditioners. Examples include S. C. Johnson & Son's Agree+ and Procter and Gamble's Pert Plus. Such products have been promoted as "green" because they eliminate the need for buying two separate products—thus, less packaging. Clairol has even introduced a three-in-one product that combines shampoo and conditioner with something called a protectant.

But all of these companies could learn a lesson from Dr. Emanuel H. Bronner, creator of Bronner's Pure-Castile Soap. Sold since 1948, it promotes itself as an *18-in-1* product. Among its many uses: shaving soap, massaging soap, deodorant, facial pack, breath freshener, insect repellant, fruit and vegetable cleaner—oh yes, and shampoo and conditioner. (Recipes for all these and more are on the label.) If that's not enough, the label doubles as something of a broadside, jam-packed with quotes, advice, and principles for living. A quart bottle sells for about $7 at most natural food stores and some supermarkets; it is also available in gallon sizes. Because it must be diluted—by as much as 100-to-1 for most of its uses—a single quart can suffice for months.

phosphate-free laundry and dish detergents were invented.

Also of concern are cosmetics' health effects, which aren't entirely known. Cosmetics makers aren't required to test their ingredients for safety before a product is sold. The Cosmetic, Toiletry, and Fragrance Association set up a program in 1976 to test the safety of chemicals used in cosmetics. But that information was never made public. The industry's *Cosmetic Ingredient Dictionary* lists 884 substances that are also listed as toxic chemicals by the National Institute of Occupational Safety and Health. The industry also has a voluntary ingredient labeling program, but the industry would prefer to list ingredients by alphabetical order rather than in descending order by volume, the way they are listed in food products.

In this respect, using products that are "all-natural" or "organic" may be somewhat better than using those that contain problematic ingredients. If the coriander, chamomile, "essential oils," and other wholesome ingredients are harvested in a responsible manner, it's likely that there's some benefit to the earth from using "green" cosmetics and body care products. But there's another dimension to this: Do natural, organic, eco-safe, non-synthetic products work any better than those that contain synthetic ingredients?

The short answer is no. "My pat answer to that is poison ivy is a natural ingredient," says Dr. Marianne O'Donoghue, an associate professor of dermatology at Rush Presbyterian-St. Luke's Medical Center in Oak Brook, Ill. "Natural ingredients aren't necessarily better."

Don Davis says "non-natural ingredients have been developed over 20 or 30 years of work on these products. The work done on chemical ingredients far surpasses whatever effectiveness studies have been done on natural ingredients. Apart from their image, which is pristine, good for your skin and hair, some of the standard cosmetic products work far better that the so-called natural products. Some of the natural ingredients are very exotic—cucumber oil, things that no one has ever proved do anything. They're in there just because they prove in the consumers' mind that they have a nice clean image."

One argument against mainstream products is their use of petroleum products, which may clog the skin's pores. "Petrochemicals are actually dead dinosaur bones that have been compressed," says Heidi Schechter, a spokesperson for Aveda. "If you remove it from the earth, it's not something that the earth can replenish. We can pick a flower and plant the same flower again. Plants and flowers are very symbiotic with us—they breathe what we give off."

Dermatologist O'Donoghue believes these products cause no problem. "Petrolatum, or Vaseline, is the ideal moisturizer for hands and feet. It's too heavy for use on the face. It's an emollient, which leaves a film on the surface of the skin, which means that when the person's skin is about to lose its natural moisture, it's trapped. It's very hydrating."

On other additives, however, O'Donoghue says, "The two major ingredients that cause trouble are fragrances and preservatives." Those who are allergic have a reaction when using a product, she says, should write down the ingredients and compare them with other products.

Needs, Wants, Reactions

So the question comes down to individual needs, wants, and physical reactions. "It's a personal preference," says Stacy Fox, The Body Shop's spokesperson. "Customers want to buy naturally based cosmetics. They like to buy things that make them feel good, things that smell good, things that look good."

Heidi Catano, spokesperson for Tom's of Maine, explains it this way: "Nature provides simple solutions. If you don't need saccharin in your toothpaste, if you don't need alcohol in your mouthwash, if you don't need petroleum in your deodorant, then why use them? We feel that a lot of chemicals and synthetic ingredients aren't necessary." Tom's sells one of the only commercial toothpastes made without saccharin. Catano says that's what Tom's is about: providing alternatives for those who want them. "Tom's isn't a product for everyone. It's a product for people that are concerned about what they put in or on their bodies, for people who want to support companies that are just attempting to do things in new ways."

Can you tell if something is natural by looking at the label? No, maintains Suzanne Grayson, publisher of the *Grayson Report*, a cosmetics industry newsletter. "You could make a product feel like aloe that doesn't contain a drop of aloe. There's no way for a consumer to determine the natural content of a product."

Though makers of toiletries are under no obligation to list any ingredients but the active ones, reading labels may pay off. "Very often the cosmetics marketers achieve what the natural product enthusiasts want by including one or two ingredients," Davis says. A shampoo, for example, could be a standard formulation, Davis explains, to which a manufacturer has simply added honey and almond oil, "which probably doesn't do much in terms of getting your hair clean."

Another problem with natural ingredients is their natural tendency to break down. "If they're going to hold up on the shelf for more than six months they must have a preservative in them," says Davis. "Preservatives are made from organic compounds that have no counterpart in nature. In large amounts they might be toxic," he says, but in most products they constitute less than 3 percent.

Nancy Connell, of the New York State Attorney General's office, says, "What we need are national standards so consumers know what something means when it says recyclable or biodegradable or whatever. Standards would go a long way toward clearing this up."

Without clear standards, Philip Dickey, of the Washington Toxics Coalition, says to be wary. "I don't place a lot of credence in a lot of green claims that are being put out in other areas. So I don't have any reason to be less skeptical about personal care products."

CRUELTY-FREE PRODUCTS

The idea of "cruelty-free" personal care products is a relatively new one, and a confusing one. On the one hand, today's consumers want products that are safe to use, that won't cause irritations or other harmful reactions. On the other hand, there is growing concern over the treatment of laboratory animals used in testing these products.

Millions of laboratory animals—mice, rats, dogs, monkeys, and others—sacrifice their health, and often their lives, in the name of "science," testing a wide range of cosmetics and personal care products. Before it reaches your supermarket or drugstore shelves, every "new" or "improved" product goes through a battery of animal tests to make sure they are safe for human use. In the process, some 14 million animals die each year. All told, about 70 million animals are put through some kind of testing, many of which result in their painful disfigurement.

The nature of many of these tests can be brutal. Test animals are routinely burned, injected with poisonous substances, artificially stressed, infected with disease, and administered electric shocks. In most cases, the tests go on for days or weeks; animals are rarely administered painkillers. One notorious experiment, the Draize Eye-Irritancy Test used in the cosmetic industry to measure the irritancy of potential new products, involves putting albino rabbits in restraining devices, then administering a few drops of the test substance into their eyes. The all-too-frequent results: the animals' eyes swell and redden until they go blind. There is also a Draize Skin-Irritancy Test. Still another test, the LD-50 (short for "lethal dose, 50 percent"), involves force feeding animals with chemicals to determine how much is required to kill half of the test animals. This test alone kills 4 million animals a year, according to the **Humane Society of the United States** (2100 L St. NW, Washington, DC 20037; 202-452-1100), one of the leading animal rights groups. Before they die, animals generally experience bleeding from the eyes, nose, and mouth; an inability to breathe; vomiting; convulsions; and paralysis.

Unfortunately, not all of these animals die for the cause of vital new products such as new cancer drugs, AIDS cures, or even baldness cures. Millions of these animals die for the cause of beauty—to create such things as mascara, shampoo, mouthwash, lipstick, hand lotion, face cream, and perfume. Even worse, according to the Humane Society, such experiments yield little to no useful information about these products' potential health risks to humans. Moreover, say animal protection activists, alternatives are available that require no animal testing at all, save time and money, and provide better information to protect human health.

Cruelty-Free Companies

Despite increasingly loud protests from animal rights groups and concerned citizens, most of the leading cosmetics companies continue to use animal testing. But a growing number of small- and mid-sized cosmetics companies do not. These cruelty-free companies either rely on methods that test animals in humane ways, or do not use animal tests at all. When you purchase products from these companies, you help prevent needless animal suffering.

Keep in mind that all of these companies and products do not necessarily meet other standards important to Green Consumers. For example, many of these products—cosmetics in particular—are over-packaged, usually in plastic. Many of these firms, however, use natural products, often minimizing allergic and other reactions in users.

Companies That Don't Test on Animals

NOT TESTED ON ANIMALS

NO ANIMAL INGREDIENTS

The lists beginning on the following page was excerpted from information compiled by **People for the Ethical Treatment of Animals** (PETA, P.O. Box 42516, Washington, DC 20015; 301-770-7382), a leading activist animal rights group. As you will see, it includes many familiar brands of cosmetics and personal care products. This is not the complete list of cruelty-free companies. To obtain the complete guide, send $4.95 to PETA and ask for *Shopping Guide for a Caring Consumer*.

PLEDGED TO THE HUMANE SOCIETY OF THE U.S. • NOT ANIMAL TESTED •

The Beautiful Choice™

When shopping for cosmetics and other products, look for cruelty-free logos from PETA (above left) and the Humane Society of the United States (above right), which attest to a product's lack of animal testing, lack of animal ingredients, or both.

Companies that Do Not Test on Animals

Baby Products
The Body Shop, Inc.
Crabtree & Evelyn
Dr. Bronner's All-One Products Company
Kiss My Face (children's products)
Neutrogena Corporation
Tom's of Maine

Cosmetics
AllerCare, Inc.
Almay
Amway Corporation
Avon
Bennetton
The Body Shop
Bonne Bell
Christian Dior
Clinique
Color Me Beautiful
Estee Lauder
Freeman Corporation
Kiss My Face
Rachel Perry, Inc.
Redken Laboratories, Inc.
Revlon

Dandruff Shampoos
Amway Corporation
Avon Corporation
Neutrogena Corporation
Nexxus
Redken Laboratories, Inc.

Dental Hygiene
Amway Corporation
The Body Shop, Inc
Crabtree & Evelyn
Tom's of Maine

Fragrances for Men
Abercrombie & Fitch
Amway Corporation
Avon Corporation
Bennetton Cosmetics Corp.
The Body Shop
Camp Beverly Hills
Clinique Laboratories
Crabtree & Evelyn
Dr. Bronner's "All-One" Products Company
Giorgio
Liz Claiborne Cosmetics

Fragrances for Women
Amway Corporation
Avon
Bennetton Cosmetics Corp.
The Body Shop
Camp Beverly Hills
Christian Dior
Clinique Laboratories, Inc.
Crabtree & Evelyn
Dr. Bronner's "All-One" Products Company
Gucci Parfums
Jean Nate
Liz Claiborne Cosmetics
Redken Laboratories, Inc.

Hair Care
AFM Enterprises
Amway Corporation
Aubrey Organics
Avon
Beauty Without Cruelty Cosmetics
The Body Shop
Clinique Laboratories, Inc.
Crabtree & Evelyn

Dr. Bronner's "All-One" Products Company
Ecco Bella
Estee Lauder
John Paul Mitchell Systems
Kiss My Face
Nature's Plus
Neutrogen Corporation
Nexxus Products Co.
Nu-Skin International
Redken Laboratories, Inc.
Revlon, Inc.
Tom's of Maine

Hair Coloring

Farmavita
Ecco Bella
Redken Laboratories, Inc.

Household Products

Allens Naturally
Amway
Crabtree & Evelyn
Park-Rand Enterprises
Shaklee US, Inc.

Hypoallergenic Skin Care

Avon Corporation
Beauty Without Cruelty
The Body Shop
Clinique
Kiss My Face
Neutrogena Corporation
Redken Laboratories, Inc.

Nail Care

Avon
Barbizon International
Bennetton Cosmetics Corp.
Clinique Laboratories, Inc.
Estee Lauder Inc.
Revlon, Inc.

Permanents

Nexxus
Redken Laboratories, Inc.

Razors

Abercrombie & Fitch
American Safety Razor
Crabtree & Evelyn
Norelco

Shaving Products

Abercrombie & Fitch
Amway Corporation
The Body Shop
Clinique Laboratories, Inc.
Crabtree & Evelyn
Estee Lauder, Inc.
Gucci Parfums
Kiss My Face
Redken Laboratories, Inc.
Tom's of Maine

Skin Care for Women

Almay
Avon Corporation
Banana Boat Sun Care
Barbizon International
Beauty Without Cruelty Cosmetics
The Body Shop
Clinique
Crabtree & Evelyn
Estée Lauder
Neutrogena Corporation
Nexxus Products Company
St. Ives Laboratories, Inc.

Skin Care for Men

Abercrombie & Fitch
Avon Corporation
Beauty Without Cruelty Cosmetics
The Body Shop
Clinique Laboratories

Estee Lauder Inc.
Neutrogena Corporation
Nexxus Products Company
Redken Laboratories, Inc.

Sun Care

Almay
Avon Corporation
Banana Boat Sun Care
The Body Shop
Bonne Belle
Clinique
Estee Lauder
Kiss My Face
Neutrogena Corporation
Nu Skin International
Rachel Perry, Inc.

Toiletries

Abercrombie & Fitch
Amway Corporation
Avon Corporation
Bennetton Cosmetics Corporation
Bill Blass (Men's)
The Body Shop
Christian Dior
Clearly Natural Products
Clinique Laboratories
Crabtree & Evelyn, Ltd.
Dr. Bronner's "All-One" Products Company
Estée Lauder
Jean Nate
Kiss My Face
Liz Claiborne Cosmetics
Neutrogena Corporation
Nexxus
Redken Laboratories, Inc.
Revlon Inc.
St. Ives Laboratories, Inc.
Tom's of Maine
Virginia's Secret

Toothbrushes

The Body Shop, Inc.
Crabtree & Evelyn

Mail-Order Companies

Cruelty-free cosmetics and other personal-care products can be found in many department and cosmetics stores. Another good source are mail-order firms specializing in these types of products. Here is a select list of companies that feature cruelty-free and environment-friendly products by mail:

• **Basically Natural** (109 E. G St., Brunswick, MD 21716; 301-834-7923) offers a selection of all-vegetarian, non-animal-tested personal care, herbal hair care and cosmetics.

• **The Compassionate Consumer** (P.O. Box 27, Jericho, NY 11753; 718-445-4134) distributes a variety of cruelty-free products that don't contain animal ingredients, including perfumes, skin care products, hair care products, and cosmetics.

• **Don't Be Cruel, Inc.** (P.O. Box 46504, Chicago, IL 60646; 312-774-4144) offers a full line of cruelty-free products, including facial

Companies That Do Test on Animals

The following companies do conduct animal testing, according to People for the Ethical Treatment of Animals. Selected brand names are included in parentheses.

Alberto-Culver Co. (VO5, Tresemme)
Allgan, Inc.
American Cyanimid Co. (Old Spice)
Aziza
Bausch & Lomb
BethCo. Frag.
Block Drug Co. (Polident, Tegrin)
Boyle-Midway (Easy-Off, Woolite)
Bristol-Myers Squibb Co. (Keri, Ban)
Calvin Klein (Obsession, Eternity)
Carter-Wallace (Arrid, Lady's Choice)
Chanel
Chesebrough-Ponds (Cutex, Vaseline)
Church and Dwight (Arm & Hammer)
Clarion
Clairol Inc. (Sea Breeze)
Colgate-Palmolive, Co.
Cosmair (Cacharel, Biotherm)
Coty (Stetson)
Cover Girl
DowBrands (Saran Wrap, Fantastik)
Drackett Products Co. (Drano, Endust, Vanish)
Eli Lilly & Co.
Elizabeth Arden
Erno Lazlo
Faberge
Gillette (Braun, Papermate)
Helena Rubenstein
Helene Curtis Industries (Finesse)
Jheri Redding (Conair)
Jhirmack
Johnson & Johnson (BandAid, Tylenol)
S. C. Johnson & Son (Agree, Raid)
Johnson Products Co.
Kimberly-Clark Group (Kleenex, Huggies)
Lamaur (Apple Pectin)
Lancôme
Lever Brothers (Caress, Dove, Close-up)
L'Oreal (Ralph Lauren, Drakkar Noir)
Max Factor
Mennen Co.
Naturelle
Nina Ricci
Noxell Corp (Noxema, Raintree)
Pantene
Parfums International (White Shoulders)
Pfizer (Coty, Ben Gay, Visine)
Playtex Corp.
Prince Matchabelli
Procter & Gamble (Crest, Tide)
Reckitt & Colman (Airwick, Stick-Ups)
Richardson-Vicks (Clearasil, Bain de Soleil)
Sally Hansen
Schering Plough (Coppertone, Maybelline)
Schick
Scott Paper Co.
Shiseido Co. Ltd.
SmithKline Beecham (Contac, Oxy-5)
Sterling Drug (Bayer, Lysol, Midol)
Vidal Sassoon
Warner Lambert (Certs, Listerine, Rolaids)
Westwood Pharmaceuticals
Whitehall Laboratories (Anacin)

and body-care products, hair care products, and cosmetics. Many are biodegradable, naturally derived, chemical-free, phosphate-free, and aluminum-free.

• **Earthen Joys** (1412 11th St., Astoria, OR 97103; 503-325-0426, 800-451-4540) sells bath products and skin cleansers.

• **Ecco Bella** (125 Pompton Plains Crossroad, Wayne, NJ 07470; 800-322-9344) sells cruelty-free products ranging from cosmetics, moisturizing cream, facial cleansers, and skin toners to suntan lotion, massage oils, and perfumes.

• **Humane Alternative Products** (8 Hutchins St., Concord, NH 03301; 603-224-1361) distributes a variety of cruelty-free products, including hair care, skin care, cosmetics, perfumes, and colognes.

• **Little Red's World** (720 Greenwich St., Ste. 7K, New York, NY 10014; 212-807-0452) sells a variety of personal care products, including the Weleda line of toothpaste, soaps, moisturizing creams, and bath oils. There is also a selection of baby products.

• **Natural World Inc.** (Glenbrok Industrial Park, 652 Glenbrook Rd., Stamford, CT 06906; 203-356-0000) offers a complete line of environmentally responsible, cruelty-free products, featuring the "Skin Basics Systems," specially formulated for different skin types.

• **Sunrise Lane Products** (780 Greenwich St., New York, NY 10014; 212-242-7014) sells products for baby care, mouth care, healing, sun care, skin care, hair care, and cosmetics.

• **Walnut Acres** (Penns Creek, PA 17862; 717-837-0601) sells its own line of organics products, made from natural vegetable sources with no perfumes, artificial colorings, or preservatives. Products range from soap and shampoo to deodorant and moisturizing cream.

FOOD & GROCERIES

The local supermarket has increasingly become an exciting but confusing place. The excitement comes from the never-ending stream of new products, each with some new ingredient, package, or marketing claim that sets it off from the others. It's this very set of circumstances that makes the supermarket confusing, too.

Product manufacturers—whether of food or nonfood items—have become increasingly daring in their sales pitches. They'll boldly claim a product has no cholesterol when it is well known by nutritionists that such a product, while indeed containing no cholesterol, may be very high in saturated fat, which has been strongly linked to heart disease. So the claim, though technically correct, is quite misleading to someone concerned about heart disease. "Natural" and "light" are among the other trendy words that have come to be overused—and abused—by product manufacturers hoping to catch a wave of consumer interest.

As green products are of increasing interest to consumers, some manufacturers have become quick to label a product "biodegradable," "environment friendly," or some other term, when that isn't necessarily the case. And there are some perfectly green products out there about which little is said in regard to their environmental benefits.

Green Packaging, Green Products, Green Grocers

What are the big issues at the supermarket, environmentally speaking? For Green Consumers, there are two principal concerns:

• **Packaging:** We've already mentioned some of the key issues related to packaging and overpackaging. Nowhere is this issue greater than at the supermarket, where we buy more "packaged goods" (as they are known in the industry) than anywhere else. This is the source of the most plastic, nonreturnable, and nonessentials—the vast majority of products destined to become household waste. All of which we pay for, when we purchase these products and when we throw them away. Packaging issues relate both to food and nonfood products.

• **Products:** Some products, particularly nonfood products, that contain ingredients harmful to the environment. A few—some cleaners, for example—contaminate the air and water as they are used. Other products do their damage when they are thrown away. In most cases, there are environmentally preferable alternatives.

Beyond these issues is the issue of the supermarkets themselves: How much do their policies and practices contribute to environmental problems? Some stores, recognizing that they themselves produce a great deal of trash (cardboard boxes, spoiled food, etc.) have taken a hard look at the way they conduct business. These stores' managers recognize that being environmentally responsible is not only good for the Earth, it is also good for business. Some Green Consumers are going out of their way to patronize such stores—and to avoid stores that are not taking positive green action. See box at right for more information about what makes a supermarket "green."

Someday, choosing green products at the supermarket will be much easier than it is today. American stores will provide shoppers with some of the same environmental amenities offered by their counterparts in Europe: green products, green shelf labeling, and green product labeling. That day will come sooner if Green Consumers raise their voices and demand this information.

For now, let's look at some things you should know about food and groceries—and what you can do today to make your shopping trip a little greener.

The Green Supermarket

What makes a supermarket green? In consultation with industry experts, we've identified twelve factors on which to judge a store's environmental practices. For more information, consult *The Green Consumer Supermarket Guide* (Penguin Books, 1991).

1. Environmental Policy. The supermarket chain should have a written environmental policy stating its intentions and goals on environmental matters.

2. Consumer Information. Stores should offer environmental shopping information to consumers, helping them to make "green" choices.

3. Shelf Labeling. Stores should provide reliable shelf labels to help consumers find products that are responsibly packaged or that have other desirable attributes.

4. Shopping Bags. Stores should give customers a choice between paper and plastic shopping bags, and should take both types of bags back from customers for reuse or recycling. Also, stores should sell reusable bags to consumers at a reasonable price.

5. Recycling Drop-off. In communities that lack curbside pickup or centralized recycling facilities, stores should offer places to deposit cans, bottles, newspapers, and other recyclables. (However, in communities with curbside recycling programs, such supermarket efforts may be counterproductive to local collection efforts.)

6. Meat Departments. Supermarket meat departments or butcher shops should offer paper wrapping as an alternative to plastic packages. In addition, meat scraps and bones, instead of being discarded, should be used by the stores for rendering, soup stocks, sauces, or composting.

7. Bulk Purchases. Supermarkets should encourage the purchasing of bulk grains and other foods, as well as case-lot quantities of packaged goods.

8. Organics. Stores should offer organic, transitional, or pesticide-free produce.

9. CFC Recovery. Stores should use systems that recover and recycle ozone-damaging chlorofluorocarbons (CFCs) in their refrigeration, freezer, and air conditioning systems.

10. Supplier Conditions. Supermarkets should encourage suppliers to reduce or eliminate unnecessary packaging, or to use packaging that can be recycled or reused.

11. Internal Recycling. Stores should recycle their own cardboard, glass, aluminum, plastic, and other materials. Food scraps should be composted.

12. Energy Conservation. Stores should actively conserve water, energy, petroleum, and other resources within their facilities and vehicles.

HOW SAFE IS OUR FOOD?

We used to take food safety for granted. Only when there was some kind of scare—often in the form of a food-tainting episode created by "terrorists"—did we worry about what was in our food.

Perhaps we should have been a bit more worried. Despite federal government claims that Americans enjoy "the safest food supply in the world," there is evidence that our dinner plates contain healthy servings of some rather undesirable side dishes. The evidence has not escaped American consumers, who are increasingly questioning food safety. When the Food Marketing Institute asked consumers, "Is the food in your supermarket safe?" fully 75 percent weren't sure. Says the trade publication *Supermarket Business*: "The evidence suggests that food safety is poised to become one of the hottest public health issues of the 1990s."

Consider pesticides, for example:

- American farmers use 1.5 billion pounds of pesticides each year—about five pounds for every man, woman, and child. These pesticides end up in about half the foods we eat.
- Only about 1 percent of all food shipments are tested for pesticides. Indeed, standard government laboratory tests aren't even able to detect more than half of the 500 or so pesticides currently in agricultural use.
- According to the Environmental Protection Agency, 66 pesticides sprayed on food crops contain cancer-causing agents. Of the 560 million pounds of herbicides and fungicides used by American farmers annually, EPA says that 375 million pounds probably or possibly cause cancer.
- According to the National Academy of Sciences, only 10 percent of the 35,000 pesticides introduced since 1945 have been tested for their effects on people.
- Even worse, some 40 percent of pesticides are used to make food look good, according to a study by the California Public Interest Research Group.
- Some of the pesticides-in-food problem comes from outside U.S. borders: Some imported foods contain residues of dangerous pesticides that have been banned in the United States but are still used in other countries; one-fourth of all fresh fruit eaten by Americans comes from abroad.

• Perhaps most mind-boggling of all is that, despite this rampant use of pesticides, the National Academy of Sciences, considered the nation's preeminent body of scientists, reported in 1989 that farmers who apply little or no chemicals to crops are usually as productive as those who use pesticides and synthetic fertilizers.

Those particularly at risk from pesticides are children. According to a comprehensive two-year study by the Natural Resources Defense Council, pesticide tolerance levels established by the federal government are set only for healthy adults. Children face special danger: Because they weigh less they eat proportionally more pesticide-containing foods than adults. They also eat more fruit, which makes up an estimated one-third of preschoolers' diets, compared to one-fifth of adults'. Moreover, the younger a child is, the more susceptible he or she is to carcinogens. NRDC concluded that children will suffer significant additional cancer and brain damage from pesticide residues unless the Environmental Protection Agency tolerance levels are adjusted to account for children's consumption patterns and increased susceptibility.

In addition, the nitrogen-based fertilizers most commonly used to grow food help to worsen global warming, according to a study released in 1989 by the Marine Biological Laboratory in Woods Hole, Massachusetts, and the University of New Hampshire. The fertilizers are said to interfere with the ability of soil microbes to remove methane, one of the most potent greenhouse gases, from the air. The study pointed out that bacteria living in soil treated with nitrogen fertilizer take up far less methane gas, upon which they can feed, than do microbes in soil that has not been treated. While carbon dioxide is far more concentrated in the atmosphere than methane, methane gas is twenty times more powerful at trapping heat close to the earth's surface. Methane levels have increased by about 1 percent a year over the past decade. (See "Seven Environmental Problems You Can Do Something About.")

And then there are the effects of pesticides on the water supply. Irrigation, which accounts for nearly 70 percent of groundwater use, can carry pesticides and fertilizers into rivers, lakes, and streams. It is estimated that up to 21 million pounds of pesticides reach groundwater or surface water before degrading. Environmental Protection Agency groundwater monitoring studies have found that at least 17 pesticides have been detected in groundwater in 23 states as a result of agricultural practices.

Pesticides and fertilizers aren't the only added ingredients affecting our food and our health. Another issue is agricultural drugs. Somewhere between 20,000 and 30,000 animal drugs are used in U.S. farming, including antibiotics, hormonal drugs, and other pharmaceuticals designed to create bigger, fatter, and faster-growing livestock. Residues of some of these drugs end up in milk, eggs, and meat.

Still another big problem is bacteria. According to the Centers for Disease Control, food-borne bacteria kill approximately 9,000 Americans each year. Millions of others suffer major or minor intestinal problems as a result of these germs. Much of this results from germs that grow in the intestines of animals that have become immune to germ-killing antibiotics. Another potent bacteria produces salmonella, which, according to the U.S. Department of Agriculture, may be present in as many as one out of every three chickens sold in the United States. But the Agriculture Department itself is part of the problem: In recent years, it has turned an increasing part of the food-inspection process over to the food industry itself. In effect, the inspectors are on company payrolls.

The sum total of all this has created an agricultural industry that is producing more and more food, but at a price. The price is paid for in the effect some of these practices have on our nation's farmland. The aggressive factory-farming techniques—which use drugs and pesticides as means of increasing production—have wreaked havoc on America's topsoil, the lifeblood of our agricultural economy, and on farmland in general. Indeed, some experts have said that today's high-tech farming methods will produce tomorrow's ecological disaster.

As the cost of farming has increased over the years, farmers have had to do more with less. They have had to increase crop yields and minimize crop losses with fewer people. They've relied more on machines, removing the hands-on aspect of farming, and on pesticides. Instead of planting a variety of crops, which causes less depletion of soil nutrients, they focus on a single crop, fueled by an increasing amount of artificial fertilizers. Most of these fertilizers are made in part from natural gas, a nonrenewable fossil fuel. It has been said that using these fertilizers is much like pouring a bunch of logs on a fire: the fire will burn brightly, then wither and die. So, too, with farming. The artificial fertilizers yield magnificent crops,

An Apple a Day . . .

You may recall the Alar scare of 1989, when public concern arose over the use of the pesticide daminozide (marketed under the brand name Alar) on red apples. But Alar is only one of many problem chemicals, according to the U.S. Public Interest Research Group, a Washington, DC–based consumer group. USPIRG compiled a list of 23 other pesticides and agricultural chemicals used in apple farming that are suspected of containing cancer-causing substances:

benomyl	o-phenylphenol
captafol	parathion
captan	permethrin
chlordimeform	phosmet
dicofol	pronamide
dinoseb	silvez
folpet	simazine
lead arsenate	tetrachlorvinphos
mancozeb	thiophanate-methyl
maneb	toxaphene
methomyl	zineb
metiram	

but they also deplete the soil. And so, the following year, *more* fertilizers are needed, and the cycle continues. Conventional corn production results in the loss of an average of 20 tons of soil *per acre*.

In the end, the land is worked so hard with these techniques that it isn't able to replenish itself. Much of the soil is blown or washed away by rains faster than new soil is able to replenish it. Since the mid-1970s, U.S. farms have lost an average of 1.7 billion tons of soil every year. Our farmland is simply being swept away by unsustainable farming techniques.

The New American Farmers. A variety of less environmentally harmful farming methods are being employed by America's farmers. Not all of these methods are new; some have been around for decades. They go by a variety of names. "Low-impact farming," for example, involves a selective use of pesticides. "Crop rotation," which used to be standard practice, involves changing crop types regularly and discourages the reappearance of the same pests. And organic farming, which used to be an activity limited to a relatively small group of "health-food hippies" dedicated to getting "back to the land," has entered the agricultural mainstream, as a growing number of farmers have learned how to produce healthy crop yields without synthetic fertilizers and pesticides. Other farming techniques have been known as "biodynamic farming," "biological farming," and "natural farming."

Whatever the name, the key word among today's forward-thinking farmers is "sustainability": the ability to produce crops without hurting the land. And the evidence shows that such farming techniques are good for both the land and its people. As the National Academy of Sciences reported in 1989, "Well-managed alternative farms use less synthetic chemical fertilizers, pesticides, and antibiotics without necessarily decreasing, and in some cases, increasing per-acre crop yields and the productivity of livestock systems." The academy recommended that the federal government revise its farm policy to encourage alternative farming methods that preserve the soil and reduce the use of chemicals.

Such statements have helped turn sustainable agriculture into more than simply a trendy alternative-lifestyles phenomenon. According to the **International Alliance for Sustainable Agriculture** (1701 University Ave. SE, Minneapolis, MN 55414; 612-331-1099), institutional researchers such as universities and food marketers have paid increasing attention—often in the form of research money—to organic farming methods. Several universities, including the University of California, Michigan State University, the University of Wisconsin, North Carolina State University, and Cornell University, have developed alternative agriculture research and extension programs.

Consumers are responding as well. According to a 1989 Harris Poll conducted by *Organic Gardening* magazine, when asked "Would you buy organically grown fruits and vegetables if they cost the same as other fruits and vegetables?" 84 percent of those polled answered yes. When that same 84 percent were asked "Would you still buy them if they cost more?" over half said yes.

ORGANIC FOODS

Despite the various types of alternative farming, by the time produce reaches grocery shelves it is often labeled under the same name: organic. Unfortunately, "organic" is a much used and often abused term. Technically, *all* food is organic, in that its growth results from biological and chemical processes—even if those processes are stoked by tons of chemical fertilizers and pesticides. At the grocery store, however, the term "organic" has generally come to refer to foods grown without these chemicals.

While the precise definition of "organic" is still somewhat murky, some definitions have been accepted by a growing consensus. The following definition, for example, was agreed upon in 1989 by representatives of both organic and conventional agriculture in a meeting sponsored by the United Fresh Fruit and Vegetable Association, a trade group:

1. Organic food production systems are based on farm management practices that replenish and maintain soil fertility by providing optimal conditions for soil biological activity.
2. Organic food is food that has been determined by an independent third party certification program to be produced in accordance with a nationally approved list of materials and practices.
3. Organic food is documented and verifiable by an accurate and comprehensive record of the production and handling system.
4. Only nationally approved materials have been used on the land and crops for at least three years prior to harvest.
5. Organic food has been grown, harvested, preserved, processed, stored, transported, and marketed in accordance with a nationally approved list of materials and practices.
6. Organic food meets all state and federal regulations governing the safety and quality of the food supply.

Is organic food really more healthful than nonorganic food? There is much debate on this. Skeptics note that an apple is an apple, no matter what coaxed it out of a tree. The nutritional value of the organic apple (or any other fruit or vegetable) will be virtually identical to the nonorganic one. Still, some nutritionists do claim that organically grown food has superior nutritional value.

Nutrition aside, it is the lack of toxic substances that may give organically grown food an advantage over most commercial produce. With the continuing sagas of contaminants in food—Alar in apples, aldicarb in watermelons, and EDB in grain, among others—organically grown produce, if not more nutritious, is less risky than other produce.

What's less risky for humans is probably less risky for the environment, at least in this case. When you buy organic foods, you are buying from farmers committed to the sustainability of the environment. And that's what being a Green Consumer is all about.

Organic Certification. So, how can you make sure that something truly is "organic"? A growing number of states—including California, Colorado, Iowa, Kansas, Maine, Minnesota, Montana, Nebraska, New Hampshire, New Mexico, North Dakota, Oregon, Texas, Vermont, Washington, and Wisconsin—have enacted organic labeling laws, most of which list the farming methods that can be used in order to be certified organic.

Another source of information is product labels certifying that a particular product meets certain criteria set by a professional organization. There are four principal organizations:

- **California Certified Organic Farmers** (303 Potrero St., Santa Cruz, CA 95060; 408-423-2263) calls itself "an organization of farmers and supporting members working to promote and verify organic farming practices, and supporting all efforts for a healthier, sustainable agricultural system."
- **Farm Verified Organic** (R.R. 1, Box 49A, Medina, ND 38467; 701-486-3578) describes itself as "an internationally accepted farm-to-table product-guarantee program that determines the authenticity of organically grown foods."
- **Organic Crop Improvement Association** (3185 Township Rd. 179, Belefontaine, OH 43311; 513-592-4983) calls itself "an internationally recognized, brand-neutral, farmer-owned crop improvement, quality assurance program, which is backed by an independent third-party certification and audit-controlled system."

• **Organic Growers and Buyers Association** (1405 Silver Lake Rd., New Brighton, MN 55112; 612-636-7933) is a longtime organic certifier, which prides itself on "commitment to quality standards."

Each group has its own seal, which can appear only on products meeting each organization's standards. Those standards tend to be strict. All produce must be raised using acceptable products and techniques, and violators are subject to having their entire *field* disqualified for a number of years. Such requirements point up the longevity that many chemicals have in the soil; the life of these chemicals does not end when the produce is harvested.

In addition to these groups, many other growers set their own standards, considering themselves to be "self-certified." Some of these growers' standards are even more strict than the organizations' listed above; other growers' standards are less strict.

The **Organic Foods Production Association of North America** (OFPANA, P.O. Box 1078, Greenfield, MA 01301; 413-774-7511) is a trade organization for the organic farming industry. Members include growers, processors, wholesalers, distributors, and retailers. OFPANA will supply its "Organic Farmers Associations Council" list of organic associations in the U.S. (send a self-addressed, stamped envelope). You may then locate the organizations in your area to determine who produces organic food locally, where it is sold, and what the standards are.

Another good source for information on organic programs, growers, and activities in your area is your state's agriculture department.

Still another organization actively involved in promoting healthy, organically grown food is **Americans for Safe Food** (ASF), a project of the Center for Science in the Public Interest (1875 Connecticut Ave. NW, Washington, DC 2009; 202-332-9110), a respected consumer organization dealing primarily with health issues. ASF's book *Safe Food* (Living Planet Press, 1992) is a comprehensive guide to "eating wisely in a risky world." ASF also publishes a list of reliable mail-order companies that will ship their

Thought for Food

Here are some things to keep in mind about organic produce:

- Organic produce isn't always as attractive as nonorganic produce. But any imperfections don't affect taste or quality. The nature of organic farming is that while produce isn't always uniform and blemish-free, it can be tastier and better for you.
- Organically grown food tends to be locally grown, using less energy and creating less pollution, compared with produce shipped from long distances.
- Organic certification does not mean that a product is entirely free of chemical residues. Air pollution, drifting sprays from nearby farms, and persistent soil toxins can all contribute small amounts of pollutants.

organic products directly to individual consumers, usually via United Parcel Service; most do not require minimum orders. To obtain the most recent list, write to Americans for Safe Food at the address above; include $1.50 and a stamped, self-addressed legal-size envelope with 45 cents postage. To obtain price information, write or call each company for a complete product listing.

ALONG THE SHELVES

Here is an aisle-by-aisle tour of the supermarket shelves, with advice on things to look for and avoid.

Beverages

We drink a lot of beverages. In fact, we drink more soda than water—about 47 gallons of soda a year for every man, woman, and child, compared with about 37 gallons of plain tap water. Add juices, coffee, tea, and all of the exotic blends that seem to be cropping up on shelves each week and suddenly the beverage section has become a major shopping experience by itself.

What's the Problem? Along with all these new products have also come new types of packaging. It's no longer just bottles and cans. There are pouches and boxes and bags, some of which are made of several layers of materials that cannot easily be recycled.

• We've already talked about the best and worst types of packaging. When it comes to beverages, you can find examples at both ends of the spectrum. At one end is the aluminum can or glass bottle—a readily recyclable container in a "closed-loop" system, meaning that a can may be turned into another can, and a bottle into another bottle.

• At the other end, there are such products as Wylers' Big Squeeze Fruit Drink, consisting of six eight-ounce bottles made of several layers of different plastics, with snap-off plastic tops and paper labels in a cardboard container—all of which is shrink-wrapped in yet more plastic! A similarly packaged product, Squeezit, was named 1989's Beverage Packaging of the Year by *Food and Drug Packaging*, a trade magazine—demonstrating just how out of touch the food industry often is with Green Consumers.

Juice Box Jive. Another award-winning beverage container is the aseptic package, also known as the juice box. You can find them everywhere—those little rectangular boxes with their built-in straws. They're lightweight, convenient, and nearly indestructible.

Juice boxes are made of three layers of materials: paperboard, polyethylene plastic, and aluminum foil. It's this thin combination that gives juice boxes their special lightweight convenience. Unfortunately, these same properties make it extremely difficult to recycle juice boxes.

If all we intended to do with our trash was to send it to landfills, juice boxes would probably be a good idea, because they take up very little space when crushed. But as most Green Consumers know, landfilling is not the goal. Nearly every city, county, and state has announced plans to recycle a growing amount of our garbage, reducing the need for landfills and getting the most use out of all of our resources.

Juice boxes have been recycled in very small quantities, in programs subsidized by juice box manufacturers. They can be turned into a type of plastic "lumber," among other things. But such recycling will not be available to many people for years, if at all. So despite industry's claim, juice boxes are, for all intents and purposes, not recyclable.

Fortunately, there are substitutes for juice boxes. Several juice and drink products now come in small aluminum or steel cans, or glass bottles, all of which are readily recyclable. But buying lots of small cans or bottles isn't particularly green. The best bet is to invest in a Thermos or a set of small reusable plastic beverage containers. They can be sent to school or work—and brought home again and refilled, over and over.

One smart company is taking on the juice-box folks by offering an all-aluminum can with a built-in straw. Juice Bowl Blasters, from Juice Bowl Products, offer the best of juice boxes plus recyclability. Now, there's less reason than ever to buy aseptic packages. Juice Bowl's advertising slogan says it all: "Get out of the box! Get into the can!" For more information contact Juice Bowl Products Inc., P.O. Box 1048, Lakeland, FL 33802; 813-665-5515.

What to Look For

• Choose products made from 100 percent recyclable material. In most cases, that's glass and aluminum, although in some communities it may also include PET or HDPE plastic bottles.

• Look for products that don't have extra layers or plastic "yokes" (which hold together six-packs of bottles or cans) or plastic layers of shrink wrapping. If you do buy packages with yokes, be sure to cut them up into small pieces before discarding; marine animals have been known to get caught in the yokes' loops.

• Buy the largest-size container of a product as possible, avoiding juice boxes and other smaller containers. If you or your kids prefer small portable containers, buy a set of plastic bottles that can be used over and over. (If you can't find them in your supermarket or kitchen supply store, check with a store that sells camping equipment. Backpackers and other campers swear by plastic containers with tight, leakproof lids.)

• Stay away from powdered beverages that come in individual packets. In most cases, such products—cocoa mix, for example—are also available in larger jars or cans. (True, you'll have to mix your own cocoa, but how do the manufacturers know exactly how chocolatey you like your cocoa anyway?)

• If your community recycles PET plastic, stay clear of colored plastic bottles unless you are sure they will be accepted by your recycler. (Of course, if you can't recycle PET plastic, stick with aluminum or glass containers.)

• Look for products from local bottlers. They require the least amount of energy and pollution to get to market.

Bottled Water

Once the beverage of a small but dedicated corps of purists, bottled water is now the beverage of choice for millions of Americans. We drink nearly 2 billion gallons a year, at an average cost of about $1.15 per gallon (compared with an average price of $1.28 per thousand gallons of tap water—that makes bottled water nearly 900 times more expensive). And our reasons for drinking bottled water—whether the generic kind that comes in a plastic jug or the more expensive name-brand varieties—contain equal portions of fact, fantasy, and fashion. But to understand them requires an explanation of the three basic types of water on the market:

- **Processed or purified water:** By removing minerals, sterilizing the water through a process called ozonation, and replacing the minerals, bottlers are capable of creating waters that rival Mother Nature's finest in purity. Processed water, often made simply from city tap water, has long constituted the largest share of the bottled water market. Giant national companies like **Foremost-McKesson** and **Borden's** have the lion's share of this market.
- **Natural still water:** Also known as "mineral water," this category includes many of the popular domestic and imported brands. Still waters are commonly distributed in five-gallon bottles suitable for use with office coolers or with a simple plastic hand pump for the home. Beware, however: Some waters sold in five-gallon containers are purified, not natural waters. Natural still waters are extracted from underground springs; some are bottled at the source, while others are transported by truck or rail for bottling. Popular brands are Evian, Mountain Valley, and Deer Park, as well as a few hundred lesser-known regional brands.
- **Natural sparkling water:** These are among the priciest bottled waters and include such popular brands as Perrier, Apollinaris, and Poland Spring Sparkling Water. What makes these waters sparkle is carbon dioxide.

The distinction between "natural" and "purified" waters is at times rather murky. Even "natural" waters are sometimes "purified" through ozonation, which involves injecting heavy oxygen molecules into the water to kill bacteria. Some sparkling waters are actually natural still waters with carbonation added, and there are squabbles within the industry whether this procedure should deprive the water of its designation as "natural." Poland Spring water is extracted from a source in the White Mountains in Maine, but its carbonation is trucked in from a well in Colorado.

All of this is made more confusing by the fact that there are few laws to require that bottled water be any purer than the stuff that comes out of your kitchen faucet. While the bottled water industry may claim that its product "is a most highly regulated and monitored drinking water supply," the regulations and monitoring still don't deal with some serious problems. The regulations enforced by the federal government set standards for bottled water that are exactly the same as for tap water—and those tap water standards are not exceptional. The rules set "tolerance" levels for some very potent chemicals, but they don't specify any limits for a wide variety of synthetic compounds called "organics"—an alphabet soup of carcinogens (which cause cancer) and mutagens (which cause birth defects) like PBB, PVC, THM, and many pesticides. About 700 organics have been measured in drinking water so far; many of them have never been tested for toxicity. True, bottled water must come from protected sources (although that can include a local faucet), be bottled in facilities regulated as food plants, be processed using manufacturing practices approved by the federal government, and provide adequate labeling. But bottlers can abide by all these rules and still not even test their water for organics.

The purification and distillation processes used in making bottled waters do kill bacteria. But during the distillation process, when water is vaporized so that minerals can be removed, some of the deadly organics remain in the water vapor and get reconstituted right back along with everything else. In areas where polluted tap water is the basis of bottled water, there's an excellent chance that the bottled version will be just as polluted as the original.

Even waters from faraway mountain streams don't escape pollution. "Nonpoint" sources of pollution—where rainwater collects pollutants and sends them into streams and rivers—put chemicals into the water cycle, and they never seem to leave. Small towns that have no industry and are miles from cities have found distressing levels of pollutants in their water supplies. Product claims about product purity and cleanliness may not hold much credibility in the 1990s, when even the rain is dirty. And while many bottlers are quick to push impressive chemical analyses at you, you will be hard-

Eating to Save the Earth

Rainforest In Your Kitchen (Island Press, $10.95) is a short but fascinating examination of the links between what we eat and the extinction of thousands of species of plants and animals. This slim, 112-page sparsely written paperback attempts to make a case that by slightly modifying how we shop, eat, and garden, "we can collectively influence the operating decisions of today's corporate agribusiness and help preserve our precious genetic resources."

That's a tall order for such a small book, and author Martin Teitel handles it ably. For example, in just two short paragraphs, he makes a persuasive case for buying brown eggs instead of white. Inside, the eggs are virtually identical, he says, but "white eggs represent a genetic disaster in the making." The problem is that more than 90% of the 3 billion eggs sold in the U.S. each year come from one kind of chicken; in fact, virtually all commercial egg-laying chickens come from just 9 hatchery sources. "What would happen to our egg supply if some pestilence wiped out that breed?" he asks. Talk about putting all your eggs in one basket!

Teitel maintains that "even if a moderate percentage of shoppers switched to brown eggs, the market would respond with a shift in our national egg-producing gene pool from one breed to two." That's still less than ideal, he says, but it's a good start.

That's a sampling of what this book offers: a frightening real look at the impacts of our diets. Perhaps we'll be frightened enough to do something about it.

pressed to find one in which the bottler has analyzed the water's level of organics.

Is Bottled Water "Green"? What about the environmental impact of bottled water? Unfortunately, from that perspective bottled waters don't look good. For one thing, the resources used in packaging, and the energy costs used in shipping water around the country and overseas, are tremendous, given that most of us have adequate resources at hand. While there are no available figures on the energy costs (and associated pollution) resulting from this activity, you can be sure that it isn't insignificant.

And then there's the trash. Plastic and glass beverage bottles account for the second largest source of solid waste, and bottled water packaging contributes to that pile of nonbiodegradable trash.

So, bottled water is not a particularly green product, no matter how many pollutants you feel you may be avoiding by drinking it.

Still, if you *know* your water is unhealthy, bottled water may at least give you peace of mind. If you must drink it, buy water that is from sources nearby, minimizing transportation energy costs, and is packaged in recyclable containers such as glass. Your best bet may be to rent a dispenser that holds recyclable five-gallon glass jugs of local water. There are many home-delivery services that will bring refillable containers of water to your house on a weekly basis; check the Yellow Pages for listings.

Home Purification Products. Ever since we discovered clean water in a bottle, we've been trying to find ways of getting around paying a dollar or so per quart for it. But the alternative may cause more problems than it solves.

Most home water filters are not very effective. In fact, some have been found to *add* pollutants to water. The problem is caused by the activated carbon used in the filtering process of many systems. Bacteria stick to the carbon—as they are supposed to—but they also grow and multiply, especially overnight and other times when the tap isn't being used for a long period. Turning on the water releases the bacteria into your glass. To counter the problem, some manufacturers add metallic silver to their filters, which effectively retards bacterial growth. Unfortunately, it also adds a new contaminant to the water: high levels of silver can pose their own health problems.

Besides activated carbon filters, home water treatment devices use other methods to cleanse water: reverse osmosis, ion exchanging, and distillation, among others. Prices range from about $100 to over $1,000. It is important to keep in mind that all of these systems must be well maintained to work properly. Failure to do this may result in additional pollutants being added to the water.

The bottom line is that most filters don't screen out the really dangerous pollutants for which they were intended, most important, the organics, which are being found increasingly in water systems around the country. There are no federal regulations governing home water treatment devices, leaving it up to you to determine whether the water that comes out of the filtering device is actually better than the water that goes in.

If you are thinking about a home water treatment system, consider the following suggestions:

- **Get your water tested by a local service.** They are usually

Washout

The bottled-water industry is mopping up following 1991 findings by the Food and Drug Administration that the purity of its products leaves a lot to be desired. Among the findings: 31 percent of samples tested were tainted with bacteria. This, for water that sells for 200 to 1,000 times more than tap water.

The fact is, tap water turns out to be the source for about a third of the bottled water sold in the United States.

It's not that most tap water is all that unhealthy. A few years back, we compared a name-brand imported bottled water with tap water that originated in the Potomac River outside Washington, DC. Despite the notoriety of the barely swimmable Potomac, the two waters turned out to be relatively the same, nutritionally speaking. The pricey bottled product had a higher calcium content, but Potomac water was richer in magnesium and potassium and contained less sodium, which is linked to high blood pressure.

Oh, yes, the bottled stuff also had bubbles.

listed in the Yellow Pages under "Environmental Services." This will help you decide whether you need a filter in the first place. You may be surprised to find that your water is in pretty good shape.

• **Read up on the system you are considering.** Make sure you understand what pollutants it removes, how often you must change the filter, and whether you can do it yourself.

• **Ask for a no-obligation 30-day trial.** Many companies will let you do this so that you can see for yourself whether the cost of the device is worth it.

Coffee

Coffee has grown from a simple product into a high-status and high-tech item. Not very long ago, the choice was simply between ground and instant. Ground coffee, more often than not, was made in a percolator; instant coffee was, and still is, a matter of adding boiling water to a teaspoonful or two. How things have changed!

Today, making a cup of coffee is partly a matter of choosing the right "technology"—drip, automatic drip, plunger, or good old-fashioned percolator, among others. Concern over caffeine consumption has led to a rise in decaffeinated coffees, some of which use chemicals similar to those used in dry cleaning to remove caffeine from coffee beans. There is no information available on the environmental impact of such chemicals. (There are alternative processes available to decaffeinate coffee beans, notably water-processing.)

The way coffee is packaged is another matter. While ground coffee traditionally comes in a recyclable steel can, several brands, emulating the way coffee is sold in gourmet shops, package grounds in vacuum-sealed, foil-like pouches. These pouches, however, are composites, made of several layers of materials, and are not recyclable. (On the other hand, some supermarkets sell coffee grounds or beans in paper pouches, an environmentally desirable package.)

Even worse are all-in-one filter packs, individual packages intended to eliminate measuring and pouring grounds into a coffee filter. These are needless and wasteful.

Some coffee comes from countries that are losing their forests at a rapid rate, although the links between coffee production and deforestation aren't yet clear enough to recommend cutting down on brands from particular countries. The problems are most likely to occur in Brazil, Central America, and West Africa.

What to Look For

- Buy ground coffee in a steel can or paper bag.
- Buy instant coffee in a steel can or glass jar.
- Avoid premeasured coffee packs.
- If you buy decaffeinated, look for the water-processed variety.

Juices

Unfortunately, aseptic packages seem to have taken over this aisle; even worse, many juice boxes are also wrapped in plastic. Still, there are many juice products packaged in glass bottles and steel or aluminum cans. Some brands come as both juice boxes and as bottles or cans. When buying juice, fresh and canned are better than frozen. The amount of energy it takes to freeze and ship frozen juice is substantial. Ounce for ounce, getting frozen orange juice to market takes four times the energy of providing fresh organges, according to the Worldwatch Institute.

Powdered Mixes

Powdered drink mixes have become popular items, and in at least one way they are ecological: By eliminating water, they reduce the weight that needs to be transported from manufacturer to market. Why ship a pot's worth of tea when a couple of tea bags will allow you to make the product yourself at home?

Unfortunately, many of these mixes are packaged poorly. In striving for convenience—companies apparently believe we prefer only premeasured and individually wrapped packages—manufacturers have created some packaging monsters. Consider Crystal Light, a product of Kraft General Foods: The mix comes packaged in a plastic tub with a foil cover. The tubs are inside are high-density polyethylene plastic container with a plastic lid and a plastic safety seal. There are four or six tubs in every product. Need we say more?

There are simpler versions of many of these products, including simple, recyclable glass jars.

Sodas

We go through a lot of soda bottles—some 2.5 million plastic soda bottles every hour, according to one estimate. Indeed, soda bottles have become a symbol of our throwaway society. They litter the highways, they crowd our landfills. The good news is that many of these same containers are being viewed as one of the great environmental success stories of the 20th century. The fact that in just two short decades we have come to recycle some 55 percent of aluminum beverage cans is remarkable; much of that growth has taken place in just the past three years. There have been similar, albeit more modest gains with glass bottles.

Plastic bottles, however, do not share in this good news. Only about a third of plastic soda bottles were recycled in 1991. (See "The Problems of Plastic" for more on this.) That may change, as more and more communities collect these bottles, and more plastic recyclers accept bottles.

What to Look For

- Look for containers that are readily recyclable in your community. In most cases, that will be glass or aluminum; in some communities, it will also include PET plastic bottles.

• Avoid extra packaging, such as plastic shrink-wrapping or yokes around six-packs. Look for bottles and cans packaged in recycled cardboard containers.

• Avoid products that have foam labels. A simple paper label is all you need.

Dairy Products

There's nothing more natural than a herd of dairy cows grazing on lush green grass. Unfortunately, the modern dairy industry is anything but natural—it's a modern milk factory. Close to 98 percent of all the milk we drink comes from these factory farms, where what are advertised as contented cows are in fact treated as four-legged milk machines.

Dairy technology has become complex and high-tech, aimed at getting maximum milk for minimum cost. Cows are given antibiotics, hormones, and tranquilizers. Residues of some of these ingredients can be found in the milk itself. According to the experts, the concentrations of these residues are so weak as to be harmless.

Milk production and distribution isn't completely harmless to the environment. The factory-farm uses vast amounts of water and energy. But because milk is produced locally, transportation is relatively minimal.

The clear glass milk bottle has all but disappeared, as has home delivery of milk and dairy products. (Not completely, though; you can still find milkmen—and milkwomen—in many small towns.) Most of us buy milk in paperboard cartons or plastic jugs. While most people would assume that paper is better than plastic, this may be one case where plastic wins out. The reason: Paper milk cartons are not recyclable. They are destined for landfills and incinerators. Plastic jugs, made from high-density polyethylene (HDPE—#2 in the plastic industry coding system), are being recycled in limited numbers. It is also a limited form of recycling: Government regulations do not yet allow recycled jugs to be made into more milk jugs, or any other type of food container. However, some recycled HDPE is being used to make plastic jugs for laundry detergent and other nonfood products.

What about other dairy products? For the most part, yogurt, cottage cheese, sour cream, and other products are packaged in plastic, and there is extremely little deviation in packaging style from company to company. Cheese may be one exception. Some brands have individually wrapped slices, a convenience for those who either lack the time to slice cheese themselves, or who perhaps can't cut straight. In any case, it's an expensive waste.

What to Look For

• Look for minimally packaged products. Avoid those with individually wrapped slices and servings.

• Buy the largest quantity of a given product that you can to minimize packaging.

• Look for products packaged in cardboard instead of plastic. While all national brands we found used plastic, some local dairies still use cardboard packaging.

• Check to see if you can recycle HDPE plastic milk jugs in your community. If so, buy milk in those containers.

Frozen Foods

For many of us, answering the question "What's for dinner?" is as simple as opening the freezer door. But frozen entrees—what we used to call "TV dinners"—are far from simple when it comes to assessing their environmental impact. Not that these heavily packaged and processed products could rightfully be called "green." But accurately determining the environmental impact of frozen foods compared to other foods isn't easy to do.

The frozen-food business is hot, annually ringing up more than $20 billion worth of frozen meat, vegetables, desserts, and just about anything else that's edible. And they're not just for dinner anymore: frozen breakfasts are growing even faster, up over 20 percent a year.

Let's state the obvious: Locally grown fresh food is hands-down more ecological than processed food of any kind, including frozen. It uses little, if any, packaging and the least amount of energy to get to the marketplace. Except for food scraps, there is little waste from fresh food. Still, fresh food transported cross country and from other countries can be more energy-intensive than frozen foods, according to one study.

Having said that, let's deal with the reality that fresh foods aren't always convenient or available. And some folks can't cook, are too busy to cook, or would at least enjoy a change of pace.

Frozen food, as it turns out, fares reasonably well in energy consumption compared with canned or other packaged foods. A 1977 study of frozen versus canned peas, for example—still considered one of the better studies to date—concluded that the cradle-to-grave energy costs were equal. True, you needn't freeze canned peas all the way from warehouse to stove-top, but making tin cans turned out to be energy-intensive, too. However, the 1977 study was sponsored by the Frozen Foods Institute, an industry association. While the study has not been disputed, the jury is still out on this question.

Energy isn't the only factor. Most freezers—industrial models in warehouses, commercial types in stores, or home models—all use chlorofluorocarbons, the ozone-destroying gas. So, foods that require cold storage contribute to ozone depletion.

The many layers of packaging are another matter, of course, and it's here that frozen foods lose out to fresh foods and most other packaged ones. The ideal packaging—none at all—won't work for frozen foods. Second best would be a package made of recycled and recyclable materials. This, too, has eluded frozen-foods makers.

What hasn't eluded them is the ability to heap layers of unrecyclable plastic upon unrecyclable coated paperboard to create massive amounts of household trash. A few products we examined had as many as six layers of packaging. In most cases, few of those layers were made of materials that were either recycled, recyclable, or reusable. Despite the needless trash these products generate, the food industry seems to love these products: they are frequently given food industry awards for innovation and design.

Still another packaging issue are the health effects from microwave cooking some of these products in their packaging, as the instructions often recommend. A growing body of evidence suggests that some of the packages' chemicals "migrate" into the foods.

What to Look For

- Choose brands with the fewest layers of packaging. Ideally,

Zap!

Still another packaging concern is what happens to some packages when used in a microwave oven. There is evidence that chemicals contained in some of the most common types of microwave packaging may "migrate" into foods when exposed to the high temperatures of microwaves. The Food and Drug Administration has conducted tests which found that certain chemicals consistently leaked into food during cooking. One FDA test found that some frozen food containers intended for cooking in *conventional* ovens also leaked chemicals into food. Even the standard "cling wrap" many consumers use to cover leftovers before reheating in a microwave are suspected of migrating chemicals into foods during cooking. (Fatty foods are a particular problem: A British study found that cling wraps can leak chemicals into fatty foods at room temperature and even during refrigeration.) The bottom line is that there are no legal definitions for such terms as "microwave-safe," "microwave-approved," and even "microwaveable." Their use does not mean that the containers can be safely used in a microwave oven.

The best bet when microwave cooking is to cover the dish with glass. Wax paper works fine to cover a dish. If you must use a plastic container or wrap, don't let it touch the food.

there should be only one or two layers.

• Buy products that have reusable components, such as resealable containers or plastic dishes. But don't buy these if you don't plan to actually reuse these things. Note that some packages specifically warn, "Do Not Reuse."

• Be aware of packaging that states "Not to be used in a conventional oven," if that's where you plan to cook it.

Meat, Poultry, and Fish

The first question Green Consumers should ask is: To eat or not to eat meat. That is the question a growing number of Americans are asking. It's not that vegetarianism is sharply on the rise—it has slowly grown to include about 7.5 million vegetarian Americans. But even meat eaters are eating less meat. Moreover, there are dramatic shifts in the kinds of meat people do eat.

Beef, heavily advertised as "real" food, is on the decline. That's good. Raising most beef cattle is a highly polluting and energy-intensive endeavor. We've already mentioned the vast amount of water it takes to produce the meat for just four quarter-pound hamburgers—2,500 gallons, according to John Robbins, author of *Diet for a New America*. (All told, raising livestock consumes more than half of all the water used in the United States.) Producing a pound of beef also requires the energy equivalent of one gallon of gasoline. But water and energy use are only the beginning. Consider these facts from *Beyond Beef*, by Jeremy Rifkin:

• There are currently 1.28 billion cattle on earth, taking up 24% of the land mass of the planet and consuming enough grain to feed hundreds of millions of people.

• Since 1960, more than 25 percent of the forests of Central America have been cleared to create pasture land for grazing cattle. Each imported hamburger requires the clearing of five square meters of jungle for pasture.

• Cattle emit methane, a potent greenhouse gas that is responsible for 18 percent of the current global warming trend.

• Nearly half the water consumed in the U.S. goes to grow feed for cattle and other livestock. To produce one pound of grain-fed steak requires hundreds of gallons of water. The water that goes into a 1,000-pound steer would float a destroyer.

• Beef has the highest concentrations of herbicides of any food sold in America and represents nearly 11 percent of the total cancer risk to consumers from pesticides of all foods on the market today.

• Each pound of feed lot steak costs about 35 pounds of eroded topsoil.

• Cattle produce nearly 1 billion tons of organic waste each year. The nitrogen from the cattle waste is converted into ammonia and nitrates and leaks into ground and surface water, where it pollutes wells, rivers, and streams, contaminating drinking water and killing marine life.

There's more. Beyond the environmental consequences, there are the human costs of beef—nutritionally, politically, and sociologically. Rifkin, a longtime activist, lays out his polemic in hard-hitting style. Regardless of whether or not *Beyond Beef* leads you to give up red meat—or all meat—you'll never eat a burger or steak again without thinking.

So much for "real" food. Raising poultry minimizes some of these problems—for example, most commercial chickens don't need any grazing area because they are confined to coops—but it

Green Eggs, No Ham

Simply put, *Recipes from an Ecological Kitchen* (William Morrow, $25) is an extraordinary work: well-designed, comprehensive, informative, and very readable. This 492-page hardcover contains more than 250 original recipes developed from the point of view that "what is good for our health and satisfies our palate can be good for the earth—and easy on the pocketbook, too."

Author Lorna J. Sass, a credentialed food writer, begins with a description of the "ecological pantry"—the staples and condiments that will be the basis of much that follows—as well as an overly brief explanation of the benefits of buying organic. Without much further ado, she gets into the recipes, from soup to nuts.

You won't find any meat, fish, or eggs here—this is a vegetarian cookbook through and through. Sass sets out to dispel the myth that vegetarian cooking is tasteless, time-consuming to prepare, and heavy. Most of her recipes are low in fat and all are cholesterol-free.

The back of the book is a gem in itself: a 70-page, A-to-Z listing of how to buy, store, and prepare hundreds of ingredients. There are also eco-tips scattered throughout on energy efficiency, food storage, and other topics.

If you are concerned about your diet, and about the effect of your diet on the environment, *Recipes From an Ecological Kitchen* should sit on your bookshelf, right alongside *The Moosewood Cookbook* and *The Joy of Cooking.*

still requires considerable resources, including grain and water. According to one statistic, it takes about 400 gallons of water to produce a single serving of chicken, about one-sixth the amount needed for a hamburger patty.

Fish—and fish farming, a fast-growing source of Americans' seafood—is the least energy- and resource-intensive, although it is far from pollution-free; current farming techniques require dumping massive amounts of fish waste into waterways, which can affect other fish, plants, and organisms.

As we said, some beef eaten by Americans comes from tropical rain forests. According to the Rainforest Action Network (RAN), an environmental group, "The typical four-ounce hamburger patty represents about 55 square feet of tropical forest—a space that would statistically contain one 60-foot-tall tree; 50 saplings and seedlings representing 20 to 30 different tree species; two pounds of insects representing thousands of individuals and more than a hundred different species; a pound of mosses, fungi, and microorganisms; and a section of the feeding zone of dozens of birds, reptiles, and mammals, some of them extremely rare. Millions of individuals and thousands of species of plant and animals inhabit a patch of tropical forest destroyed for a single hamburger."

Who uses the beef from rain forests? You won't find much of it in the supermarket as ground beef or steak; beef raised in rain forest regions is said to be stringy, tough, and cheap and often goes into mass-produced foods, where it is combined with fattier domestic beef and cereal products. RAN urges consumers to avoid purchasing such processed-beef products as baby foods, canned beef, frozen beef products, hot dogs, luncheon meats, and soups.

Some companies, such as Campbell Soup Company, have stated that while they import beef from South America, none of it comes from rain forest areas. But RAN responds that, regardless of whether this is true, there are other reasons to avoid all beef imported from Central and South America: "This beef is more likely to be contaminated with toxic chemicals, trace metals, and organic contaminants than that raised in the United States. Excessive pesticide residues have repeatedly been found in beef prepared by packing plants in Costa Rica, El Salvador, Guatemala, and Mexico, for which these packers have been decertified by the U.S. Department of Agriculture."

What about fast-food hamburgers? The major chains, such as McDonald's, Burger King, and Wendy's, say they don't use rain forest beef; McDonald's has a strict company policy that states that it "does not, has not, and will not permit the destruction of tropical rain forests for our beef supply," and that "Any McDonald's supplier who is found to deviate from this policy or who cannot prove compliance with it will be immediately discontinued."

Organic Meat. The idea of meat being organic may seem contradictory at first, but it is growing in popularity. A relatively small number of organic ranchers are producing beef and poultry products that are free from the myriad of ingredients—antibiotics, pesticides, fertilizers, hormones, and other goodies—that are injected into or otherwise fed to livestock. Such techniques don't necessarily make organic meat "green"—relative to other foods, there are still significant impacts on the environment—but they are certainly "greener" than their nonorganic counterparts.

How Now?

Jeremy Rifkin clearly has a beef with beef.

His 1992 book, *Beyond Beef,* offers a powerful argument against cattle and the industries that produce and market hamburgers, steaks, and other forms of beef. Indeed, Rifkin has stated flat out: "Except for the automobile, there is no more destructive force on earth than the cow. ... Cattle literally threaten the future of the earth."

Rifkin, who has founded such organizations as the Greenhouse Crisis Foundation and the Peoples Bicentennial Commission, has launched a new group, inspired by this book: the Beyond Beef Campaign.

The campaign, supported by the Humane Society of the U.S. and other environmental and animal-rights groups, aims "to influence consumers around the world to cut their consumption of beef by at least 50%," replacing it with grains, fruits, and vegetables. For those still eating some beef, the campaign will advocate the consumption of beef "humanely raised under strict organic standards." The campaign includes TV and radio public service announcements.

Membership in the Beyond Beef Campaign is $25, which includes the *Beyond Beef Newsletter,* updates and briefing materials, and educational and organizing packets. For more information write to 1130 17th St. NW, Ste. 630, Washington, DC 20036.

For more information about organic meat companies, contact the **International Alliance for Sustainable Agriculture,** 1701 University Ave. SE, Minneapolis, MN 55414; 612-331-1099.

Tuna and Dolphins

Ten days before Earth Day 1990, the three leading U.S. tuna canners made a dramatic announcement: They would stop buying tuna that were caught in a manner that harmed dolphins. It was a bold move, one heralded by environmentalists as a major step in stemming the needless deaths of up to 150,000 dolphins a year.

For reasons unknown, tuna often gather just below herds of dolphin. So tuna fleets watch for dolphins to locate their catch. Then, in a practice known as "setting on dolphins," tuna fishermen snare the tuna—and dolphins—in plastic driftnets up to 30 miles long. Trapped, the air-breathing dolphins suffocate or drown. The dead or wounded dolphins are cast back into the sea.

The announcement—by H. J. Heinz Company, which markets

StarKist; Van Camp Seafood Company, which markets Chicken of the Sea; and Bumble Bee Seafoods Inc.—followed by passage of the Marine Mammal Protection Act, was supposed to put an end to all this. And in large part it did. That is, until Mexico sued the United States under an international treaty called the General Agreement on Tariffs and Trade (GATT), charging that the American law was a "barrier to trade"—that is, it hindered the Mexican fishing industry from selling tuna in U.S. markets. The GATT panel agreed, saying the law had to go. As of this writing, the status of the law remains in limbo.

But that doesn't mean that dolphin-safe tuna isn't widely available. It is, and Green Consumers should look for the dolphin-safe logo—or words to that effect—whenever buying canned tuna.

Pet foods are another concern. Heinz's brands, including 9 Lives, have been deemed dolphin-safe, but no others have yet to pass Earth Island Institute's standards. Like supermarket house brands, much of the pet-food tuna is caught and processed in Thailand, making verification difficult.

For additional information about tuna and dolphins, contact the **Earth Island Institute,** 300 Broadway, Ste. 28, San Francisco, CA 94133; 415-788-3666.

Packaging. As usual, packaging of meat and fish products is a concern. At the supermarket meat counter, the butcher paper of yesterday has given way to polystyrene plastic trays covered with polyethylene plastic wrap. Supermarket managers say this is necessary because, unlike the butcher shop of old times, today's customers can't inspect each cut of meat personally before it is wrapped. And plastic, they say, helps maintains meat freshness and provides sanitation on supermarket shelves.

Perhaps. But a growing number of supermarkets are responding to Green Consumer concerns by wrapping meat in paper for customers who request it. Such policies may not be announced or posted in meat departments, so you'll have to ask.

As for packaged meat and meat products, the usual rules apply: the least amount of packaging, and the greatest percentage of recycled and recyclable packaging.

What to Look For

• Buy meat, poultry, and fish wrapped in paper instead of plastic whenever possible. Ask the store's fish or meat department to make this available to all customers.

• Avoid processed meat products. Some contain rain forest beef. Others contain low-grade beef from Latin America that contains pesticides and other undesirable ingredients.

• When buying packaged meats, look for minimally packaged products. One layer of packaging should be sufficient.

• Look for dolphin-safe brands of tuna. Make sure it states this specifically on the label.

• When shopping for tuna, buy the biggest-size can possible. Avoid convenience-size packages containing three single-serving cans in a cardboard container.

Fruits and Vegetables

We've already briefly discussed some of the environmental impacts of modern farming. The fact is, we've become a nation of consumers addicted to produce perfection: Perfectly shaped and blemish-free fruits and vegetables, available all year long. That's a tall order, to be sure, and American food growers, processors, and shippers have done an admirable job of catering to our every whim.

Want some juicy red tomatoes in the middle of February? No problem. They may be shipped to you from the fields of Puerto Rico or have been sailed, flown, and trucked several thousand miles from the farm to your family's dinner table. And all that time they will be kept under refrigeration to maintain freshness. Even domestic produce can travel great distances. The average mouthful of food travels about 1,300 miles to get to your dinner table, according to the Worldwatch Institute.

Of course, a lot of fruits and vegetables don't make the entire trip in their original form. Along the way, they are canned, bottled, frozen, or otherwise processed and packaged. For example, nearly half of all the fruits Americans consume go through some processing and packaging before being purchased and consumed.

Each step of the food-distribution process—growing, processing, transporting, warehousing, and selling—requires energy and resources, and contributes to pollution. Much of these resources are necessary to getting adequate and healthful food to our tables. But some of it is not. No one is suggesting that we go back to a diet composed only of locally grown and seasonal foods, but our developing tastes for convenience and exotic foods does clearly have important implications. In our environmentally conscious world, we may need to ask ourselves whether it really makes sense to bring so much of our food from so far away.

A trip through the produce aisles yields some surprises, not just in the fruits and vegetables you'll find, but in the way some of them appear. Does a grapefruit, with its thick rind, really need to be shrink-wrapped in plastic? Certainly not, but you can find grapefruits—not to mention tomatoes, bell peppers, cauliflowers, lettuce, even bananas—constrained inside plastic wrappers, sometimes mounted on cardboard or polystyrene trays! There's simply no need for this.

And then there's the packaging for frozen, canned, and processed produce. Fortunately, the overwhelming majority of it is packaged in steel cans and glass bottles. But some of it comes in plastic, usually multi-resin plastics that cannot be recycled.

What to Look For

• Choose fresh produce whenever possible. Besides being healthier than the processed kind, it has the lowest "energy content" because it has used a minimal amount of resources to get it from farm field to dinner plate.

• Choose produce grown locally whenever possible. Ask the produce manager where you shop which fruits and vegetables come from nearby farms and which come from overseas. Make it known that you intend to avoid imported produce as much as possible.

• When selecting fresh produce, everything needn't be placed in its own plastic produce bag before being purchased. Several varieties of produce purchased by item (as opposed to by weight) can be "ganged" into one bag. At the checkout stand, the clerk will have no problem determining the price of each item.

• Ask for organic or no-pesticide produce. Besides being local, it requires fewer fertilizers and pesticides to grow. That's not only healthier for you, it's better for the environment.

Up In Smoke

Here's a chilling thought you might keep in mind before your next barbecue: Cooking meat over charcoal is polluting. We've long known that charcoal fires themselves foul the air. But, reported *Science News* in 1991, meat emits dozens of compounds as it cooks, including a bevy of hydrocarbons, furans, steroids, and pesticide residues. Researchers at California Institute of Technology found that the particles emitted from grilling are of the size most easily inhaled of all organic aerosols. The amount of pollution you emit when grilling has to do with the kind of meat you cook. Researchers found that "fine aerosol emissions from charbroiling of regular hamburger meat are higher than from charbroiling extra-lean meat, which in turn are higher than the emissions from frying the same amount of hamburger." The researchers did not compare the relative amount of pollution emitted from grilling chicken or fish, but since emissions seem to be proportionate to the amount of fat, it is likely that these meats cook more cleanly.

• When buying prepackaged fruits and vegetables, choose products packaged simply, in aluminum, steel, and glass. Avoid plastic packages.

• Consider growing your own produce. Even if you have just a small plot of land, you can grow delicious tomatoes, peppers, and other produce. It will be healthy, economical, and fun.

Snacks and Single Servings

Most of us were taught not to snack between meals, but as we've grown, times have changed. Just about any time is snack time, it seems, even mealtime. Indeed, for a lot of busy folks, meals—lunch in particular—consist of a series of snacks: a granola bar, some chips or crackers, a yogurt, candy bar, popcorn, pudding, or whatever.

Sometimes a quick meal is in order. Increasingly, many such meals include microwaveable or ready-to-eat single-serving dishes. From salad and soup to main course and dessert, there are few eat-on-the-go courses you can't find on supermarket shelves.

Snacks and convenience are part of our way of life. Busy parents can have kids' meals standing by (kids can even make them themselves), brown baggers can get beyond the standard-issue tuna or bologna sandwich at lunch, and older and disabled people can prepare their own meals where they might otherwise rely on others—or simply skip eating altogether.

But convenience has a price, both at the checkout stand and in the amount of energy consumed and trash produced through the manufacture, use, and disposal of these products.

Can we balance ease with environment? In most cases, the answer is yes, if you shop wisely.

Sometimes, a little creativity is in order. Want a microwaveable noodle soup? You can buy any of the half-dozen or so soups on the shelves, but all are packaged nearly the same: an unrecyclable plastic tub with a plastic or foil top, usually in a cardboard box, which may itself be shrink-wrapped.

Or, you could buy a package of ramen, a dry noodle product sold in most supermarkets. It comes in a single thin plastic wrapper. You'll have to use your own bowl and add water before microwave cooking, but if you've got a microwave to begin with, you've probably got access to these few simple resources. The difference between the two methods is not only a lot of trash. The overpackaged noodle soup sells for about 35 cents an ounce, compared to just 11 cents an ounce for the ramen.

There are many similar examples of ways to cut packaging and costs for snacks and convenience meals. It can be as simple as buying a large-size package, then doling out individual servings in resealable plastic containers.

What to Look For

• Choose snacks packaged in the least amount of packaging, and the highest percentage of recycled and recyclable materials. Especially avoid individually packaged single servings.

• Try to buy the largest-size package you can of a product. If you need smaller portions, divide them up yourself in reusable and resealable plastic containers. Even cookies and chips can be divided into smaller resealable "zipper" plastic bags. (The bags can be rinsed out and reused.)

• When buying cardboard boxes of cookies and crackers, look for recycled boxes. (They have a gray underside; if it's white, it is unrecycled cardboard.)

Green-Bagging It

Every new school year means another nine months of packed lunches. That, in turn, makes for a lot of trash. Even if you don't have school-age kids, you may be among the millions of workers who bring lunch from home occasionally. If so, you know firsthand the amount of paper and plastic trash left over each day at the end of a meal.

That needn't be. By reusing and recycling, you can keep a large percentage of that trash out of landfills. Here are some suggestions:

- Rather than wrapping sandwiches in foil, plastic wrap, wax paper, or sandwich bags, use washable, resealable plastic containers.
- Avoid using a fresh new paper bag every day. If you don't like the idea of folding it up and sticking it in your pocket or purse, consider using a small cloth tote. EcoSource (P.O. Box 1656, Sebastopol, CA 95473; 800-274-7040) sells a 100-percent cotton lunch bag for $5.95; even better are its "Multi-Use Eco-Sacks," washable, unbleached cotton bags, $1.25–$1.75.
- Avoid juice boxes, which are very difficult for most people to recycle. Try pouring beverages from larger bottles into small plastic containers with screw-on tops, available in many grocery stores. Better yet, use a Thermos.
- Buy large sizes of chips and other snacks, and transfer a day's portion into a reusable container. You'll also save money buying in bulk.

You might consider the "Litterless Lunch Box" series from Rubbermaid, which consists of all-in-one contraptions designed to eliminate juice boxes, sandwich bags, and other disposable trash. The line includes more than 30 reusable (and colorful) food-to-go containers, from a model designed to fit neatly inside a backpack, to a hard-hat-style version, to 4-, 5-, and 7-quart models. Each comes with resealable plastic containers for sandwiches, salads, and beverages. You can do away with drink boxes, sandwich bags, and other throwaways. Look for them at mass marketers such as Wal-Mart and Kmart, or contact Rubbermaid, Inc., 1147 Akron Rd., Wooster, OH 44691; 216-264-6464.

What about school policies that discourage or prohibit nondisposable matter in lunches? Several schools still have such policies. Try working with principals or administrators to overcome objections, whether it is safety, health, or something else.

Will having a "greener" lunch really save the Earth? Of course not. But going the disposable route starts kids off with a throwaway mentality; going the greener route will instill green values on a daily basis. Indeed, it might be the most important lesson they learn all school year.

Baby Foods

We once considered baby food as pure as the driven snow—select blends of apples, carrots, potatoes, and other fruits of the earth. Sadly, today's apples, carrots, potatoes, and other produce are no longer assumed to be either pure or safe. From concerns about residues of the pesticide Alar on apples to baby-food company disclosures of synthetic ingredients being sold as 100-percent natural ingredients, many parents have had second thoughts about the purity of some baby foods.

Consumer concerns about chemicals in food have led to the spawning of the organic baby food industry. Still a relative toddler, with less than 1 percent of the $1 billion-a-year baby food business, the organic baby food market is growing steadily as its products become more widely available. The market received a boost from a 1989 study by the Natural Resources Defense Council that warned of lifetime risks of cancer for today's schoolchildren from pesticide residues. According to the NRDC, between 5,500 and 6,200 of today's preschoolers are likely to develop cancer solely because of exposure to just eight pesticides.

There is only one principal manufacturer of organic baby food:

• **Earth's Best Baby Food** (P.O. Box 887, Middlebury, VT 05753; 802-388-7974, 800-442-4221) is available in natural food stores nationwide, and by mail order in case quantities from **Hand In Hand** (Rte. 26, R.R. 1, Box 1425, Oxford, ME 04270; 207-539-8305, 800-872-9745). It is also available through diaper services in several cities. Earth's Best also offers vegetarian dinners for babies. The line includes six dinners, including one that features whole-wheat pasta with chick peas, green beans, carrots, and basil in a light tomato sauce. Each vegetarian dinner provides 3 to 4 grams of protein per jar, which the company says is comparable or higher than the 2 to 3 grams found in traditional meat-based infant dinners. As with all the company's products, these are certified organically grown, meaning the fruits, vegetables, and cereals are grown without using any synthetic pesticides or fertilizers.

Keep in mind that making your own baby food is not difficult and is the most economical and ecological way to go. You will be sure to avoid additives and unnecessary packaging, and can personally ensure that your baby is getting a healthy, balanced diet.

What to Look For

• Choose baby foods packaged in recyclable containers, such as glass jars and steel or aluminum cans.

• Buy the largest-size product you can. Avoid single-serving containers whenever possible.

• If you are concerned about pesticides, buy organic baby foods. Better yet, make your own.

Kitchen Wraps and Bags

In many respects, kitchen wraps and bags are some of the most environmentally—and economically—important purchases we make. By effectively protecting leftovers and other stored foods, materials such as aluminum foil, plastic wrap, and wax paper save food—and money—from being thrown away. While only foil is both recyclable and contains recycled material, other materials, used sparingly, are good buys.

It's when these materials are destined for disposability that they create both opportunities and challenges for Green Consumers. One of the big problems has to do with degradable trash bags. In 1989, Mobil Chemical Co. (which makes Hefty brand trash bags) and First Brands (Glad brand bags) introduced "biodegradable" trash bags. Other companies soon introduced similar products. (Some were known as "photodegradable," others simply "degradable.") It was such a seductive notion: By adding just a few teaspoons of corn starch to the recipe for plastic bags, we could make them virtually melt away in the noonday sun.

Alas, it was not to be. While degradable under laboratory conditions, they did not simply "melt away" in a landfill, where there is little oxygen or sunlight to facilitate the degrading process. Landfills, as we've learned, tend to mummify trash—including trash bags—better than dispose of it.

It wasn't as if these degradable trash bag makers had simply miscalculated. Indeed, some of them admitted that they didn't really believe the degradability claims, but were using it as a marketing tool because that's what people were asking for. The

Recycled Trash Bags

The following trash bags have been tested by Scientific Certification Systems and certified to be made from recycled plastics. Most of these companies make several types of bags: 33-gallon trash bags, tall kitchen bags, lawn and leaf bags, etc. Included in parentheses is the total percentage recycled material, as well as the percentages of post-consumer waste (PCW) and pre-consumer recovered industrial material (PRE). PCW is plastic collected from curbside or community recycling programs; PRE is scrap material generated in the manufacturing process.

SCIENTIFIC CERTIFICATION SYSTEMS™

- Amway trash bags (80% recycled: 24% PCW, 56% PRE) (Webster Industries, 508-532-2000)
- Best Buy trash bags (80% recycled: 11% PCW, 69% PRE) (Dyna-Pak Corp., 615-762-4016)
- Full Circle trash bags, tall kitchen bags, and leaf and grass bags (80% recycled: 11% PCW, 69% PRE) (Dyna-Pak Corp., 615-762-4016)
- Harmony™ and Recycle 1™ trash bags (80% recycled: 30% PCW, 50% PRE) (North American Plastics Corp., 708-896-6200)
- Mr. Neat Bags Again™ trash bags (95% recycled: 15% PCW, 80% PRE) (Strout Plastics, 612-881-8673)
- Renew® trash bags (100% recycled: 11% PCW, 69% PRE) (Webster Industries, 508-532-2000)

Federal Trade Commission and a task force of state attorneys general took a harsh view of this behavior, and levied fines against several companies.

So, while degradability is a bogus issue, recyclability is a valid one. In recent years, plastics manufacturers have managed to find ways to make trash bags from relatively high percentages of recycled plastic. Two brands—Renew, from Webster Industries, and Ruffies Eco-Choice, from Carlisle Plastics—are made with as much as 100 percent recycled plastic, with about 30 percent coming from curbside-recycled plastic bottles.

But be careful. Not all recycled bags are worth buying. Just as with unrecycled trash bags, some recycled ones are very weak, prone to break open with only a small load. It pays to shop carefully.

What to Look For

• When buying trash bags, ignore claims of degradability. Look for bags that contain at least some recycled plastic.

• When buying kitchen wraps, buy the largest size package you can to minimize packaging.

• Buy sandwich bags made of cellulose, which is made from wood pulp, instead of plastic, which is made from petroleum.

Other Things to Do

• When covering foods before microwave cooking, use wax paper instead of plastic wrap. Wax paper allows steam to escape, reducing the chance you'll burn yourself. And when plastic wrap comes into contact with food while microwave cooking, there is a chance some of the chemicals in the wrap may leach into the food.

• Try to minimize your use of kitchen wraps by keeping stored foods in resealable and reusable plastic containers.

• Recycle as much as possible, so you'll use fewer trash bags.

PAPER PRODUCTS

Paper. What could be more environmentally responsible? It comes from natural, renewable resources and degrades easily and quickly when thrown away. How bad could a paper towel be, anyway?

Plenty bad. In fact, it could be downright deadly.

The problem with paper towels—and many other household paper products, including toilet paper, disposable diapers, coffee filters, milk cartons, tampons, and facial tissues—has to do with the presence of dioxins, a family of toxic, carcinogenic chemicals, one of which was used as a weapon by American troops in Vietnam as the deadly "Agent Orange." A growing body of research has found traces of dioxins in many consumer paper products. There is concern about the health effects that may result when these products come into contact with food or with sensitive parts of the body.

What are dioxins and how do they get into paper? "Dioxin" refers to a chemical family with 75 individual members. While all have the same basic chemical structure, some are more toxic than others. Scientists believe that dioxins imitate natural steroid hormones, such as estrogen, in our bodies and can trigger a wide range of biochemical reactions. Minute quantities can trigger anything from acne and achy joints to insomnia, cancer, birth defects, and immune system disorders. Moreover, dioxins (and a chemical cousin called "furans") tend to accumulate in the body because they are fat soluble; they are stored in the fat cells of each organism. As always, children are especially sensitive—dioxin is found in the milk of the average North American mother. It is possible for nursing infants to be exposed to up to two hundred times more dioxin than healthy adults. Based on animal tests, the EPA has classified dioxins as a "probable human carcinogen."

Dioxins are formed in pulp making during the chlorine bleaching process. The problem is most severe in the manufacturing of bleached "kraft" pulp—a process that uses a sulfide soak to cook the wood chips in. The process produces a strong, dark-colored pulp suitable as feedstock for manufacturing paper. It must then be bleached in a five- or six-stage sequence to achieve high brightness. Chlorine is used to bleach the pulp. Higher-grade papers, such as most printing papers, require fewer bleaching stages and less chlorine, resulting in less dioxin formation. Newsprint, because it is not bleached, is not known to contain dioxins.

The dioxins created in pulp making don't just end up in the paper. In fact, scientists first made the link between dioxin and paper after discovering unexpectedly high levels of the chemicals in several fish downstream of pulp and paper mills. The high dioxin levels in the fish made them unsafe for human consumption. Moreover, when dioxin-contaminated water is used to irrigate crops, they, too, become contaminated. The Environmental Protection Agency reported in 1988 that, when grown in contaminated soils, root crops—such as carrots, potatoes, and onions—can develop dioxin levels that equal or exceed those found in the soil itself.

But it is the paper products themselves that present the most immediate problem. An official at the Food and Drug Administration told *Science News* in 1989 that the two major sources of dioxins were milk cartons and coffee filters. The official's "very rough estimates" were that young children getting all their milk from contaminated cartons might double their dioxin intake. Heavy coffee drinkers consuming most of their brew from pots with bleached-paper filters might increase their daily dioxin intake 5 to 10 percent above the average U.S. level.

Recycled Paper Products

The following brands of household paper products have been tested by Scientific Certification Systems and certified to be made from recycled paper. Unless otherwise noted, each of the brands includes facial tissues, bath tissues, paper towels, and napkins. Included in parentheses is the total percentage recycled paper, as well as the percentages of post-consumer waste (PCW) and pre-consumer recovered industrial waste (PRE). PCW is paper collected from curbside or community recycling programs; PRE is scrap material generated in the manufacturing process.

SCIENTIFIC CERTIFICATION SYSTEMS™

- Aware (from 10% PCW, 90% PRE, to 18% PCW, 82% PRE) (Ashdun Industries, 201-944-2650)
- Capri, Gayety, Gentle Touch, Nature's Choice, and Pert (28% PCW, 72% PRE) (Pope & Talbot, Inc., 503-228-9161)
- C.A.R.E. (from 10% PCW, 90% PRE, to 18% PCW, 82% PRE) (Ashdun Industries, 201-944-2650)
- Cascades (100% recycled: 80% PCW, 20% PRE) (Cascades Industries, 819-363-2704)
- Doucelle bath tissues and kitchen towels (80% PCW, 20% PRE) (Cascades Industries, 819-363-2704)
- Envirocare and Enviroquest (from 10% PCW, 90% PRE, to 18% PCW, 82% PRE) (Ashdun Industries, 201-944-2650)
- Forever Green, Green Meadow, New Day's Choice, Safe, and Tree-Free (10% PCW, 90% PRE) (Statler Tissue Co., 617-395-7770)
- Green Forest bath tissues, napkins, and paper towels (10% recycled: 90% PCW, 20% PRE) (Fort Howard Corp., 414-435-8821)
- Mor-Soft bath tissues and napkins, Rose Soft bath tissues, and Morning Glory napkins (60% PCW, 40% PRE) (Morcon, Inc., 518-677-8511)
- Project Green (from 10% PCW, 90% PRE, to 18% PCW, 82% PRE) (Ashdun Industries, 201-944-2650)
- Start (18% PCW, 82% PRE) (Orchids Paper Products Co., 714-523-7881)
- Today's Choice (89% PCW, 11% PRE) (Confab Co., 714-955-2690)

But almost all bleached paper products made from virgin pulp are suspect. Tests sponsored by the paper industry found that superabsorbent disposable diapers, paper towels, tea bags, tampons, juice cartons, TV dinner containers, and various types of paper plates also have contained low levels of dioxin. Dioxins from any of these products can migrate into foods or onto sensitive parts of bodies. Moreover, when these paper products are disposed of in a landfill or are incinerated, the dioxins released in the air can be inhaled by animals and humans or ingested through contamination of food crops.

It is important to keep in mind that *the smallest detectable amounts of these compounds have been known to cause cancer in laboratory animals.*

By the way, dioxins aren't the only toxic substances found in the wastewater of pulp and paper mills. A 1986 study by the Ontario Ministry of the Environment found a total of 41 substances of concern, from aluminum to zinc. Included in that list were some of the most notorious pollutants: benzene, cadmium, lead, mercury, polychlorinated biphenyls (PCBs), toluene, and others. Ironically, some environmentalists fear that focusing public attention on dioxins might derail industry and government action on some of these other pollutants.

Avoiding Dioxins. The good news is that you can avoid the hazards of dioxins in paper products by buying unbleached or chlorine-free paper products. The bad news is that, due to lower demand, such products are not readily available in the U.S.

Part of the problem has been the paper industry's reluctance to admit, for liability reasons, that dioxin in conventionally bleached paper might be a potential health problem. Another part of the problem is the industry's belief that consumers will buy only white paper products. Yet the industry's own poll has told them that if consumers knew about dioxin contamination, they'd avoid buying bleached paper products.

Another partial solution to dioxin contamination is the use of recycled paper. Use of gaseous chlorine in pulp mills has generally been associated with the creation of dioxins. However, the bleaching processes of many recycled paper companies are not believed to create dioxins. This is because paper recyclers do not use wood pulp and bleach with a weaker form of chlorine. Moreover, heat is a significant factor in the formation of dioxins; recycled paper is manufactured at a lower temperature than virgin paper.

Alternative bleaching techniques are being used by paper mills in Europe and by some recyclers in the U.S. Oxygen, peroxide, and sodium hydroxide are being used instead of chlorine. Although the cost of converting a paper mill from chlorine to other bleaching methods is high, there may be considerable savings in the cost of bleaching chemicals—not to mention the added value of saving human health and the environment.

There is no doubt that the answer to the problem rests with the shopping choices made by Green Consumers. Paper companies will not dare change their financially successful paper recipes unless they know that you won't continue to buy their products.

You can help deliver that message by buying the recycled and unbleached products currently available. Here are several sources:

Unbleached Coffee Filters

• Permanent gold mesh coffee filters are available from kitchen supply, gourmet coffee shops, and department stores. One brand is **Swissgold**.

• Reusable cotton coffee filters can be used for up to one year. They are available from **Clothcrafters** (Elkhart Lake, WI 53020; 414-876-2112); **Earthen Joys** (1412 11th St., Astoria, OR 97103; 503-325-0426); and **Seventh Generation** (49 Hercules Dr., Colchester, VT 05446; 800-456-1177).

• Unbleached paper coffee filters are available from many coffee retailers and kitchen supply stores, and directly from the manufacturers: **Ashdun Industries** (400 Sylvan Ave., Englewood Cliffs, NJ 07632; 201-569-3600), sold under the "C.A.R.E." brand; **Melitta U.S.A.** (1401 Berlin Rd., Cherry Hill, NJ 08034; 609-428-7202, 800-451-1694); and **Natural Brew** (P.O. Box 1007, Sheboygan, WI 53082; 414-459-4160).

GREENER CLEANERS

Let's start with the bad news: There is little that is clean or green about most household cleaning products.

That holds true across the board, from Ajax and Fantastik to the newest breed of so-called "biodegradables." The fact is, almost anything we dump down our drains, even if derived from plants and other "natural" substances, can cause problems.

Making matters worse, there are few concrete definitions or legal requirements for most of the common buzzwords used on product labels—words like "nontoxic," "biodegradable," and "natural." And manufacturers aren't required to fully disclose most of their ingredients—they are considered trade secrets—and even when disclosed, the list of hard-to-pronounce chemicals (such as sodium lauryl ethoxysulfate) will mean little to most consumers.

Equally confusing is the trade-off between environmental and human safety. Some environmentally preferrable products are more harmful to people than their less ecological alternatives. Low-phosphate detergents, for example, while reducing water pollution, are 100 to 1,000 times more caustic than phosphate detergents, according to the Household Hazardous Waste Project (HHWP). This means low-phosphate detergents can cause serious burns if even a small amount is ingested.

The good (or at least better) news is that there is an increasing number of greener products on the market, though comparing one to the others may require a degree in chemistry, to say the least. And while none of these may warrant our unqualified recommendation, some offer environmental improvements over their competitors.

Breaking Down Is Hard to Do

One of the popular claims among today's alternative cleaners is that they are "biodegradable." This implies that a cleaner's ingredients break down quickly and harmlessly after they go down the drain.

Biodegradability usually refers to a cleaner's surfactants, the active agents that perform the main cleaning function in most detergents. The rate at which they break down is very important. Those that break down slowly or incompletely endanger plants, animals, and micro-organisms that live in water systems.

According to experts, many so-called biodegradable products contain surfacants that do not degrade fully or quickly enough to minimize their environmental impact. Even at low concentrations, some surfacants can increase the penetration of other harmful chemicals through the protective layers of plants and animals.

Moreover, not everything that's biodegradable is good. Take phosphates, for example. They biodegrade totally and quickly. Indeed, they're a potent source of nutrients for plants in water. When they get into rivers and lakes, they cause algae blooms, robbing the water of oxygen, blocking sunlight, and ultimately killing marine life. So biodegradability is not necessarily a panacea. Says Steve Zeitler, vice president of Chempoint Products Co., which makes cleaning products, "Biodegration doesn't mean anything because some products are harmful until they biodegrade."

In general, surfactants derived from petroleum oil fatty acids contain impurities that break down much more slowly. Surfactants made from natural vegetable oil fatty acids generally break down more quickly and completely.

But not all petrochemical-based surfacants are bad, and not all plant-based ones are good. Says Judy Amand of the Pennsylvania Resources Council, "Something that comes from a plant can be just as toxic as something that comes from a fossil fuel. Fossil fuels are just plants that have been sitting around for a long time."

This Sud's For You

Phosphates are another problem. Their presence in cleaners—primarily laundry and dishwashing detergents—softens water and increases the ability of surfacants to do their job. According to the Soap and Detergent Association, detergents with phosphates yield better performance per dollar. But as stated earlier, phosphates increase plant growth in water, killing water life. For that reason, they have been banned in several states.

When phosphate-free detergents were first introduced in the 1970s, they were not embraced by consumers, due to their poor performance. Consumers used twice as much detergent or washed loads twice, then switched back to brands containing phosphates Today's versions have been vastly improved. When *Consumer Reports* tested eight phosphate and phosphate-free brands in 1987, it found little difference between the two.

Some of the phosphate substitutes are not without problems. Zeolite, for example—used in Ecover's detergents—is relatively inert but, being insoluble, must be removed from the sedimentation process in sewage plants. Still, zeolite is among the best alternatives to phosphates for the time being.

Bleaches are also problematic. They whiten, brighten, and remove stains from fabrics. Most common is chlorine, which has the added benefit of disinfecting and deodorizing. But in wastewater, chlorine can create organo-chlorine compounds, some of which are toxic to humans. Although chlorine is substantially diluted by the time it reaches the waste stream, it is still undesirable. Non-chlorine bleaches are gentler to clothes and the environment. The most common is sodium perborate. Non-chlorine bleaches are less effective in colder-water temperatures, requiring more energy-intensive hot water.

A growing number of products—including many familiar brands sold in supermarkets—have dropped chlorine bleach, usually in favor of peroxide-based bleaches. However, one of the common peroxide bleaches, perborate, produces boron when it breaks down, which can be harmful to plants in some concentrations. Moreover, the process used to produce these bleaches—by reacting either borax or soda with hydrogen peroxide—requires considerable amounts of energy and raw materials.

So, "chlorine-free" does not necessarily mean a cleaner is environmentally benign.

Wrapping It Up

We haven't even touched upon packaging. While almost all powdered detergents come in recycled cardboard boxes, liquid detergents—increasingly the preferred choice—usually come in bottles made of polyethylene terepthalate (PET) or blends of several types of plastic. While PET is beginning to be recycled in small quantities, it is not considered to be an environmentally desirable material. Plastic blends, of course, are totally unrecyclable. Most of the "alternative" cleaners listed on the opposite page are packaged in plastic, though many are concentrates, which reduces packaging somewhat. Few seem to offer refills to their spray bottles. One firm, Earth Rite, boasts its bottles are made of "easily recyclable HDPE

Biodegradable Cleaners

The following cleaning products have been tested by Scientific Certification Systems and certified to be biodegradable—that is, that they break down into carbon dioxide, basic minerals, and water:

SCIENTIFIC CERTIFICATION SYSTEMS™

- Clear Magic automotive cleaner, household cleaner, and industrial cleaner (Blue Coral, Inc., 216-641-5490)
- Descale-It Bathroom Cleaner (Descale-It Company, 602-622-2826)
- EarthRite all-purpose cleaner, all-surface floor cleaner, counter top cleaner, dishwashing liquid, glass cleaner, liquid laundry detergent, toilet bowl cleaner, and tub and tile cleaner (Benckiser Consumer Products, 203-731-5035)
- Enforcer Drain Care Powdered Drain Cleaner (Enforcer Products, Inc., 404-386-0801)
- J. R.'s All-Purpose Cleaner Degreaser Concentrate (Eden Sales & Marketing, 602-371-0069)
- K-37 Septic Tank Treatment Bacterial Additive (Roebic Laboratories, Inc., 203-795-1283)
- Klean-Strip Pre-Paint Cleaner (W. M. Barr & Co., Inc., 901-775-0100)
- Planet all-purpose household spray, concentrated all-purpose cleaner, dishwashing liquid, and glass cleaner (Planet Products, Inc., 604-656-9436)
- 20/10 All Season Windshield Cleaner Concentrate (20/10 Products, Inc., 800-638-2623)

[high-density polyethylene]," an exaggeration at best.

Plastic spray bottles have at least one advantage: they are handy for use with homemade cleaners (see page 168).

Things to Look For

Confused? You're not alone. Nearly every expert we talked to readily admitted the confounding state of affairs and the lack of easy solutions. Making things worse, so many federal agencies are involved that no one has a handle on the situation.

What should you do? Here are some suggestions:

- Don't accept vague words like "biodegradable" or "nontoxic" at face value. Ask companies to substantiate their claims—in plain English.

• If you have questions about specific ingredients, contact HHWP or a local Poison Control Center (often listed in the front of the phone book). Most centers have data about chemicals' potential health hazards.

• Avoid problem ingredients such as phosphates and chlorine. But don't assume that a phosphate- or chlorine-free product is safe.

• Buy concentrates whenever possible. Ask manufacturers to produce refillable versions that allow you to refill a spray bottle by adding water to a packaged concentrate.

• Avoid products tested on animals. Most are made of ingredients that have been around for some time. Animal testing causes great suffering to the animals involved.

• Press for legislation to disclose toxics in household products, and to establish standards for environmental labeling claims.

How to Make Your Wash Come Out "Green"

Doing the laundry is no one's favorite chore, but there's no reason to add injury to insult by creating pollution in the process. Indeed, you might be surprised how dirty this cleansing process can be to the planet.

But not necessarily in the ways you think. Laundry detergents have received the most attention, due in part to the phosphates and other ingredients that have been known to foul our nation's rivers, lakes, and streams.

But detergents aren't necessarily the only dirty part of Wash Day. They are simply the most confusing laundry ingredient because they can contain any of a number of problematic ingredients. There are other environmental considerations of doing the laundry, not the least of which is the amount of energy used to wash, dry, and iron our clothes, and the ingredients in some of the other products we use to make our wash come out whiter, brighter, fluffier, and smelling like a sunny day in May.

So, what exactly *are* all these ingredients? Besides phosphates and bleaches, detergents have these key ingredients:

• The basic ingredient of any detergent is a surfactant—chemical shorthand for "surface active agent"—an organic chemical that, when added to a liquid, changes the properties of that liquid. Simply put, a surfactant lowers the water's surface tension, enabling the cleaning solution to better remove soil from clothes. Some surfactants also help keep dirt emulsified, suspended, and dispersed so that they are less likely to resettle on clean clothes. There are literally hundreds of surfactant chemicals; most detergents contain a mixture of several.

• Detergents also include fluorescent whiteners, also known as optical brighteners (which enhance fabric brightness by making visible ultraviolet light bouncing off of fabrics; however, the effect cannot be seen under incandescent lighting). Also found in most detergents are NTA and EDTA (sodium nitrilotriacetate and ethylene diamine tetraacetic acid, respectively, which inactivate minerals in hard water), perfumes (more than 4,000 compounds are used), and preservatives.

Some of these ingredients are thought to be harmful, but there's precious little hard evidence or agreement among the experts. Regarding EDTA and NTA, for example, "For every study you show me saying one thing, somebody else will show me another saying the opposite," says Allen Conlon, owner of Allens Naturally, which makes alternative cleaning products.

However, there's some agreement that the more ingredients present, the more complex the detergent becomes, and the more difficult it is to break down quickly and completely. The fact is, laundry detergents needn't contain so many bells and whistles. Unfortunately, determining which products are simplest isn't easy.

It's important to keep in mind that almost all commercial detergents were designed specifically to clean synthetic fabrics. Natural fibers—cotton, wool, silk, and linen, for example—can be cleaned using other, safer ingredients.

In your search for simplicity, it may be worth considering some of these time-tested techniques:

• Using a water softener or conditioner can minimize mineral deposits and soap scum that can make fabrics look dull and dingy. However, some laundry soaps already include softeners.

• If you must use a commercial bleach, choose a powdered, nonchlorine brand.

• For heavily soiled items, soak them in warm water for 30 minutes with a half-cup of washing soda.

Laundry Aids

Detergent is only the beginning. Doing the wash often involves several other products, including softeners and starches. Here are some environmentally safer, cheaper alternatives to some of these commercial products:

• **Fabric Softeners:** Most aren't very harmful, though some people are sensitive to their dyes and fragrances. The fact is, softeners are designed for synthetic fabrics; you don't need any with natural fabrics. Fabric softeners should conform to the same environmental standards applied to laundry soaps.

• **Starch:** Commercial starches frequently include formaldehyde and pentachlorophenol, both of which are suspected carcinogens. Most products are also aerosols, which are not environmentally desirable. As an alternative, dissolve two tablespoons of cornstarch in a pint of cold water in a spray bottle. Shake before each use and spray as necessary while you iron.

Power Cleaning

The greatest environmental effects of Wash Day probably have little to do with these chemicals and more to do with the energy it takes to run the washer and dryer.

Heating the water alone consumes 90% of that energy. Using hot water for both washing and rinsing uses three and a half times more energy than washing in warm water and rinsing in cold water. In fact, you should never rinse in anything but cold water. Water temperature is only important during the wash cycle. The only time you'll need hot (as opposed to warm or cold) water for washing is when dealing with greasy stains.

Of course, the machine you use plays a key role in energy use. According to the American Council for an Energy-Efficient Economy (ACEEE), a private, nonprofit organization, the most efficient standard (over 16 gallon capacity) washing machines on the market use four times less energy than the least efficient machines. That alone can save up to $70 a year in energy costs. If

Spin Cycle

If you're interested in green cleaning, you should know about the Clean Up Campaign launched by Ecover, which makes ecologically sound laundry and household cleaners. Ecover hopes to "influence the U.S. government to impose regulations on the ingredients found in household cleaning products," according to spokesperson Ellen Weiser. Its efforts are focusing on Congress, where the 1972 Clean Water Act is up for revision. Ecover is seeking language to restrict and eventually eliminate the use of detergent ingredients that are known contributors to water pollution. For more information, contact Ecover, 4 Old Mill Rd., P.O. Box 1140, Georgetown, CT 06829; 203-853-4166.

every household in the United States used the most efficient machines available, it would result in saving the equivalent of up to 40 million barrels of oil a year, according to ACEEE.

Front-loading machines are more efficient than top-loading ones, using a third less water and energy. Moreover, *Consumer Reports* found that front-loaders offered better overall washing performance. Ironically, while common in Europe and other countries, front-loaders have all but disappeared from sale in this country; only one company, White Consolidated Industries, makes them (under the White Westinghouse, Gibson, and Sears brands). See "Home Energy and Appliances" for more on energy-efficient washers and dryers.

Here are some other "green" laundry tips:

- Check your water heater. Lowering the temperature to 120° will suffice for most household needs and reduce your energy costs.
- It's most efficient to fill the washer whenever you do laundry. It takes less energy to do one big load than two smaller ones. But don't overload the machine or nothing will get cleaned.
- Check your dryer's outside vent. Make sure it is clean and closes properly, or it will allow cold air into your house.
- When shopping for appliances, check the EnergyGuide labels. They tell you how much the machine will cost to operate. Don't be distracted by the purchase price. A higher initial price tag can cost considerably less in the long run if the machine's annual operating costs are lower.
- Reduce your energy bills by using nature's dryers—air and

sun—to do the work of your dryer, at least during the summer. Wooden dryer racks that fold up can hold almost as much as a clothesline if cold weather or lack of space prevent you from using a clothesline outside.

• You can reduce the need for ironing by taking clothes out of the dryer slightly damp and hanging them up. That will save energy, including yours.

PRODUCTS FOR THE AGES

Green cleaners don't have to be new, alternative, limited-distribution products found only in health food stores and in mail-order catalogs. Some products have been around for years. Chances are good your grandparents used them, and they may be in your cupboard, too. Armed with these six relatively pure and simple products, you can clean just about anything in the house.

• **Arm & Hammer Baking Soda** (Church & Dwight Co., 469 N. Harrison St., Princeton, NJ 08543; 609-683-5900) has been around since 1846. Besides being used for baking, it is a mildly abrasive cleaner that also absorbs odors. Baking soda is derived from a naturally occurring mineral left behind after evaporation of an inland lake in Wyoming 50 million years ago. The mineral is converted and purified into sodium bicarbonate—baking soda. The components of baking soda—bicarbonate and sodium ions—are present in significant concentrations in the human body. Bicarbonate helps maintain proper acid/base balance in the blood, stomach, and saliva. Sodium plays a key role in the body's functioning. Sodium bicarbonate has been sold in the United States since the 1840s and is listed by the Food and Drug Administration as a "generally recognized as safe" substance.

Because of its unique properties and benign environmental impact, baking soda provides effective, economical, and ecological alternatives to many household cleaning chores, from removing scuff marks from linoleum floors to rinsing hairspray and shampoo buildup from hair and hairbrushes. You can even keep a box in your car's glove compartment for use as an engine fire extinguisher.

Good Things In Small Packages

When buying commercial laundry products, seek out "ultra" brands, which pack more product into less packaging. Most major powdered laundry detergent brands are available in "ultra" versions, which require less detergent per laundry load. Procter & Gamble also has introduced superconcentrated liquid detergents—"Ultra" Tide, Cheer, Era, Dash, and Ivory Snow—in refillable bottles made of at least 50% recycled plastic. P&G also makes Ultra Downy fabric softener, which also comes with its own refills.

• **Bon Ami Polishing Cleanser** (Faultless Starch/Bon Ami Co., 510 Walnut St., Kansas City, MO 64106; 816-842-1230) contains no chlorine, phosphates, dyes or perfumes. The detergent in it is biodegradable; it also contains sodium carbonate and a water conditioner; feldspar and calcite are the abrasives. Because of its mild abrasive quality, Bon Ami can be used on porcelain and stainless steel fixtures, as well as on cookware, glass-top ranges, cultured marble, fiberglass, and ceramic tile. Bon Ami can also be used to clean butcher-block tops, woks, food processors, white shoes, luggage, boats, and swimming pools.

• **Dr. Bronner's Pure Castile Soap** (All-One God Faith, Box 28, Escondido, CA 92033; 619-743-2211) is sold primarily at health food and camping supply stores. This mild soap—with its famous small-print label crammed with uses, recommendations, and advice from Dr. Bronner himself—has provided reading material on many a bookless camping trip. The soap is biodegradable and extremely versatile. The label lists 18 uses, from shaving and shampooing to treating athletes food and purifying water. Invented in 1935 by Bronner to kill the odor of diapers, it has been on the market since 1941.

• **Fels Naptha** (Dial Corp., 111 W. Clarendon Ave., Phoenix, AZ 85077; 602-207-2800) is a rugged agate-colored bar soap invented in 1894. A staple of some laundry rooms, it can also be used to help deter the effects of poison ivy, especially if you wash with it directly after exposure to the weed. Fels-Naptha is versatile: one gardener uses it as an insect repellent, shredding it and sprinkling it around plants.

• **Ivory Soap** (Procter & Gamble Co., 1 Procter & Gamble Plaza, Cincinnati, OH 45202, 513-983-1975) dates back to 1879, when a worker at Harley Procter's soap factory accidently allowed air to work into the recipe for a bar of soap. The result was a soap that floated. Originally called White Soap and later renamed Ivory, it has been the cornerstone of Procter & Gamble, now the nation's biggest consumer goods manufacturer. Ivory Soap Flakes—later, Ivory Snow—came along in 1919. Boasting a legacy of 99 $^{44}/_{100}$ purity Ivory bar soap and laundry detergent are among the mildest cleaners around, suitable for a wide range of applications.

• **20 Mule Team Borax** (Dial Corp., 111 W. Clarendon Ave., Phoenix, AZ 85077; 602-207-2800) made its first appearance in 1890. A mild abrasive, it comes from boron, a mineral mined from the earth's crust in Death Valley, California. The product comes from the large number of mules needed to haul heavy wagons full of borax over rugged terrain. Borax is a good disinfectant and mold killer; it's also a very cheap household cleaner. It can be used for killing household odors, as a polish for stainless steel, as a toilet bowl cleaner, as a fabric whitener and softener, and as a stain remover for blood, chocolate, and grease. Some people also have used it to kill household fleas by sprinkling borax on their carpet, letting it sit for a couple of hours, then vacuuming it up.

NATURAL CLEANERS

The greenest cleaners of all reside in many of the products you probably have in your kitchen: baking soda, vinegar, lemon juice, vegetable oil, borax, and good old hot water, for example. While not necessarily as alluring as a "magic-bullet" spray product that promises to clean everything under the sun in mere seconds, most of the alternative cleaning methods below are just as effective, far less expensive, and nonpolluting.

The best information source on make-it-yourself cleaners is *Clean & Green: The Complete Guide to Nontoxic and Environmentally Safe Housekeeping*, by Annie Berthold-Bond, available for $10.45 postpaid from Ceres Press, P.O. Box 87, Woodstock NY 12498.

Here are some time-tested suggestions:

Art Imitates Glade

An intriguing new product called NonScents ArtForm allows you to sculpt or mold your own air freshener. What's behind ArtForm is a natural mineral adsorbent called zeolite, which is used to filter out air pollutants and odors in hospitals, office buildings, hotels, and other places. According to ArtForm's manufacturer, the zeolite is blended with specially treated recycled paper and clays to make a white powder that is mixed with water to form a doughy clay. You can add color to the clay or paint the sculpted or molded clay after baking it in the kitchen oven. ArtForm items can be baked out to release the trapped pollutants and reused endlessly.

For more information contact the Dasun Company, P.O. Box M, Escondido, CA 92033; 800-433-8929.

Aerosols: Even though they no longer contain CFCs, the nonrecyclable containers are a waste of resources and pollute the environment; also, the microscopic aerosol-propelled particles may be harmful to the lungs, heart, and central nervous system when inhaled. Avoid aerosols; if you must spray use pump products.

Air fresheners: These mask odors by coating nasal passages and deadening nerves to diminish the sense of smell. Don't use commercial air fresheners. Instead: find sources of odors and eliminate them and keep your home and closets clean and well-ventilated. Try setting out 2 to 4 tablespoons of vinegar or baking soda in open dishes. Also consider using house plants; they are good air purifiers. Another recipe is to boil herbs and spices for natural fragrance.

All-purpose cleaners: Ammonia and chlorine are in many all-purpose cleaners. These form deadly chloramine gas when mixed. Ammonia itself can be harmful to lungs, while chlorine can form cancer-causing compounds when released into the environment. Make your own cleaner by mixing two teaspoons of borax and one teaspoon of soap in one quart of water in a rinsed-out spray bottle. Or use a half cup of washing soda (hydrated sodium carbonate) in a bucket of water; it works on all but aluminum surfaces.

Carpet deodorizers: Sprinkle baking soda or cornstarch on carpet, using approximately one cup per medium-size room. Vacuum after 30 minutes. Or mix two parts cornmeal with one part borax, sprinkle liberally, leave one hour, and vacuum.

Dishwashing liquids: Most dishwashing liquids are detergents, derived from petroleum; they are nonbiodegradable and usually contain chemical additives such as artificial fragrances and colors. (Detergents also cause more child poisonings than any other household product). Use liquid or powdered soap such as Ivory (add two to three teaspoons of vinegar for heavy soil). In dishwashers, use equal parts borax and washing soda (hydrated sodium carbonate); increase the proportion of soda for hard water.

Disinfectants: Most disinfectants are a mix of toxic chemicals including phenol, formaldehyde, cresol, ammonia, and chlorine. Instead, mix one-half cup borax in one gallon hot water. This was tested in a California hospital for one year and met all state germicidal requirements, according to the Clean Water Fund.

Drain cleaners: The lye, hydrochloric acid, and sulfuric acids in drain cleaners can burn human tissue, causing permanent damage. If not used according to instructions, they can explode; these are especially dangerous around children. Prevention is the best strategy here. Never pour grease down a drain, always use a drain sieve or hair trap, and clean the metal screen or stopper mechanism regularly. Once a week, as routine maintenance, plug the overflow drain with a wet rag, pour a quarter cup baking soda down the drain, follow with one-half cup vinegar, and close the drain tightly until fizzing stops. Flush with one gallon boiling water. For persistent clogs, use a rubber plunger or a metal drain snake.

Flea and tick control: Most pesticides used for flea and tick control have never been adequately tested for safety. These products rub off the pet onto people and furniture, exposing your family to the risk of cancer and other diseases. Feed two tablespoons of brewer's yeast and a clove of raw garlic to pets daily. Also, pet dips and sprays containing de-limonine gas derived from citrus extracts safely repel pests. Pyrethrin powders made from ground chrysanthemums sprinkled on the carpet, then vacuumed, prevent further infestation. Insecticidal soaps are biodegradable and nontoxic, and kill fleas, ticks, and lice instantly. Wash pets with warm soapy water, dry

Put This In Your Pipes

Finding effective drain-cleaning products that don't contain highly caustic and toxic ingredients hasn't been easy. But Scientific Certification Systems (SCS) gave its blessing to two such products: Enforcer Drain Care and Enforcer Septic Tank Treatment. Both are based on the technique of bioremediation, in which bacteria produce enzymes that "eat" harmful chemicals and liquidize fats, oils, greases, and household wastes that have accumulated in pipes. The bacteria are grown on 100 percent bran. SCS has certified the biodegradability of the products—that they break down safely and efficiently and that they do not build up harmful concentrations in the environment. Both products are available in hardware, grocery, and retail stores nationwide. The drain cleaner retails for about $8, the septic tank cleaner about $4, both for 16-ounce canisters. For more information contact Enforcer Products, Inc., P.O. Box 1068, Cartersville, GA 30120; 800-241-5656.

thoroughly, and use the following rinse: one-half cup fresh or dried rosemary in one quart of boiling water; steep 20 minutes, strain, and cool. Spray or sponge onto pet and allow to dry (don't towel dry). Also, organic repellents made from distillates of cedarwood, orange, eucalyptus, and bay are available for house sprays and flea collars. (See "Pets and Pet Supplies.")

Floor cleaners: Dull, greasy film on no-wax linoleum can be washed away with one-half cup white vinegar mixed into one-half gallon warm water.

Floor and furniture polish: Many wood polishes contain phenol, which causes cancer in laboratory animals. Ingesting one thimbleful of phenol can cause symptoms ranging from circulatory collapse to death. Residual vapors contaminate the home long after use. Also, wood polish may cause severe skin irritation. Mix a one-to-one ratio of vegetable oil and lemon juice or vinegar into a solution and apply a thin coat. Rub in well. On unwaxed wood, use vegetable oil and lemon oil to replenish shine.

Glass cleaner: Commercial products emit ammonia mist, which enters the lungs. Ammonia is a poison—use it only when other cleansers won't do the job. *Never mix ammonia with bleach or* *commercial cleansers—deadly fumes may result.* First, use alcohol to clean the residues left from commercial glass cleaners. Then clean the glass with a mixture of half white vinegar and half water.

Insect repellents: Commercial insect repellents contain a variety of toxic chemicals. There are several safe and effective alternatives, such as burning citronella candles. Or plant sweet basil around the patio and house to repel mosquitoes. Still another method is to blend six cloves of crushed garlic, one minced onion, and one tablespoon soap in a gallon of hot water. Let sit one or two days, strain, and apply with a spray bottle.

Laundry products: Most laundry powders are nonbiodegradable detergents. Use phosphate-free, biodegradable detergents, or better yet, switch to soap flakes (such as Ivory). When switching, wash items once with washing soda (hydrated sodium carbonate) only. This eliminates detergent residues that might react with soap to yellow fabrics. Boost soap products with washing soda; this will brighten all washable fabrics and costs less than bleach.

Metal polishes: The fumes from phosphoric and sulfuric acids and ammonia contained in commercial metal polishes contribute unnecessarily to indoor air pollution. For silver, soak 10 to 15 minutes in one quart warm water, one teaspoon baking soda, one teaspoon salt, and a small piece of aluminum foil; wipe with a soft cloth. Or rub with a paste of baking soda and water. For aluminum, dip cloth in lemon juice and rinse with warm water; or soak overnight in a mixture of vinegar and water, then rub. For brass, mix equal parts of salt and flour with a little vinegar, then rub. For chrome, rub with undiluted vinegar. For copper, rub with a paste of lemon juice or vinegar, salt, and flour, or hot vinegar and salt. For gold, wash in lukewarm soapy water, dry, and polish with a chamois cloth.

Mold and mildew cleaners: These may contain pesticides. Instead, make a concentrated solution of borax or vinegar and water, and clean affected areas. For mildew, try a mixture of lemon juice or white vinegar and salt. Borax will inhibit mold growth.

Mothballs: These often contain p-dichlorobenzene, a known carcinogen. Use cedar chips or herbal sachets; store woolens in

Using Cleaning Products Safely

Here are some suggestions for using and storing household cleaning products safely, excerpted from *Guide to Hazardous Products Around the Home, 2nd Edition*, available for $8 postpaid from the Household Hazardous Waste Project (1031 E. Battlefield St., Ste. 214, Springfield, MO 65807; 417-889-5000).

• Read all labels carefully before using hazardous products. Be aware of their uses and dangers.

• Leave products in their original containers with the label that clearly identifies the contents. Never put hazardous products in food or beverage containers.

• Do not mix products unless instructed to do so by label directions. This can cause explosive or poisonous chemical reactions. Even different brands of the same product may contain incompatible ingredients.

• If you are pregnant, avoid toxic chemical exposure as much as possible. Many toxic products have not been tested for their effects on unborn children.

• Avoid wearing soft contact lenses when working with solvents and pesticides. They can absorb vapors from the air and hold the chemicals near your eyes.

• Use products in well-ventilated areas to avoid inhaling fumes. Work outdoors whenever possible. When working indoors, open windows and use an exhaust fan, making sure air is exiting outside rather than being circulated indoors.

• Do not eat, drink, or smoke while using hazardous products. Traces of hazardous chemicals can be carried from hand to mouth. Smoking can start a fire if the product is flammable.

• Make sure containers are kept dry to prevent corrosion. If product container begins to corrode, place it in a plastic bucket with a lid and clearly label the outside container with contents and appropriate warnings.

• Store volatile chemicals or products that warn of vapors or fumes in a well-ventilated area, out of reach of children and pets.

• Store rags used with flammable products (including furniture stripper, paint remover, and gasoline) in a sealed container.

• Store products away from heat, sparks, flames, or sources of ignition. This is especially important with flammable products.

• Store gasoline only in approved containers, away from all sources of heat, flame, or sparks, in a well-ventilated area.

cedar-lined closet or trunk. Moth eggs can be destroyed by running the item through a hot dryer. (Be careful, however: a hot dryer can damage or shrink some clothes. If you are uncertain, read the washing instructions on the clothing label.)

Oven cleaners: The basic ingredient in oven cleaners is lye, a powerful caustic that can burn and disfigure. Exposure can scar your lungs or cause blindness if splashed on eyes. Prevent the need for oven cleaning by avoiding overfilling pans, and scrape up spills as soon as food is cool enough to handle. When cleaning, remove remnants of charred spills with a nonmetallic bristle brush. Clean oven with a paste of baking soda, salt, and hot water. Or sprinkle with dry baking soda and scrub with a damp cloth after 5 minutes. (Don't let baking soda touch wires or heating elements.)

Pesticides: These contain some of the most toxic chemicals around. Many have been linked to birth defects, leukemia, and cancer. As a general rule, keep kitchen, floors, and garbage pails clean to eliminate pests' food supplies and remove clutter to eliminate nesting areas. For ants, sprinkle cream of tartar, red chili powder, dried peppermint, or boric acid where they enter. For cockroaches and silverfish, use equal parts baking soda and powdered sugar. (The sugar attracts them, the baking soda kills them.) For fleas, use flea combs or herbal flea powders on pet and keep house thoroughly vacuumed. For slugs and snails, place copper-sheeting barriers around sensitive plants. For houseflies, use sticky untreated flypaper; or make your own with honey and yellow paper. For mice and rats, use mousetraps or mix one part plaster of Paris with one part flour and some sugar and cocoa powder; sprinkle where rodents (but not children) will find it. As for spiders, leave them alone—they eat other insect pests.

Toilet cleaners: These contain chlorine and hydrochloric acid, which can burn skin and eyes. Instead, use soap and borax; remove stubborn rings and lime buildup with white vinegar. Another cleaner is baking soda sprinkled into the bowl; drizzle with vinegar and scour with a toilet brush.

GARDEN & PET SUPPLIES

There is nothing more "green" than growing a garden. Whether it is a window box hanging from an apartment window or an acre behind your house, gardens provide a host of benefits to your psyche and your refrigerator. At least 69 million Americans did some type of gardening and lawn care during 1988, according to the National Gardening Association, and the numbers are growing each year. According to a 1987 Louis Harris poll, gardening is the number-one recreational activity in the United States. And all that gardening uses a lot of chemicals: a 1980 study by the National Academy of Sciences showed that residential lawns and gardens received as much as 10 pounds of chemicals per acre—compared with about 2 pounds per acre for soybean crops.

Unlike many of our recreational activities, this one is good for the environment. All growing vegetation absorbs carbon dioxide, which otherwise is released into the atmosphere and is a major contributor to greenhouse warming (see "Seven Environmental Problems You Can Do Something About"). The more vegetation that exists, the less carbon dioxide is released into the atmosphere. And, of course, if you grow vegetables, the more you grow the less you'll have to buy, alleviating your contributing to energy costs and pollution associated with most modern agriculture.

But gardening also can be far from green, depending on how it is done. Although gardening supply and chemical companies have introduced a lengthy list of products to make gardening more productive and less work, many of these products are not good for you, your children, your pets, and other living things. Even if you aren't growing anything edible—by humans, at least—you may still be harming other creatures, some of whose presence could be protecting your garden from a variety of less-than-welcome pests. The fact is, there is a variety of good-quality products and techniques you can use to have both a green thumb and a green environment.

PROBLEMS WITH PESTICIDES

Most gardeners face a dilemma: Do you deploy an arsenal of chemicals on your plants, or do you go "chemical free" and hope for the best? In the past, most people opted for the chemicals. And there have been many to choose from—an estimated 83 brands of pesticides offering between 300 and 400 individual products are sold on retail shelves of gardening, hardware, and grocery stores, as well as by mail order—for which we spend nearly $1 billion a year, according to the National Gardening Association. None of which includes the billions of pounds of pesticides dumped on crops by American farmers (see "How Safe Is Our Food?"). According to the Environmental Protection Agency, suburban lawns and gardens probably receive the heaviest applications of pesticides of any land in the United States.

The fact is, the safety of a substantial number of these products has yet to be established or already has been seriously questioned. A 1972 federal law gave to the EPA the task of re-registering all pesticides then on the market. The re-registration process includes a detailed examination of data on safety as well as both short-term (acute) and long-term (chronic) health effects. But due to weak enforcement and bureaucratic entanglement, only about 120 of the 600 principal ingredients in commercially available pesticides have been registered so far. So, don't assume that because a product is available in your local hardware or garden store, it has undergone rigorous scrutiny by the government. There's an 80 percent chance that it hasn't. Moreover, some pesticides that were once widely used have now been banned or severely restricted, including DDT, chlordane, aldrin, heptachlor, dieldrin, lindane, silvex, and 2,4,5-T. If you are storing pesticides that you haven't used in two or more years, there's a chance they might contain one or more of these ingredients.

What are the effects of all these chemicals on you and on the environment? That the effects can be deadly has been known for years—since 1962, when Rachel Carson published her landmark book *Silent Spring*, which exposed to the general public for the first time the horrid effects of pesticides on water, soil, and air, and the wildlife and people they support. As Carson put it: "For the first time in the history of the world, every human being is now subjected to contact with dangerous chemicals, from the moment of conception until death." But since *Silent Spring*, things haven't necessarily improved; in many ways they've gotten worse. There are greater amounts of a greater number of chemicals being used than ever before.

The problems aren't limited to a few well-publicized toxic waste dumps. The pesticides you use at home—and which are used in parks, golf courses, and other nonagricultural settings—are carried by the wind and are transported through the soil and sewer systems into rivers, lakes, and streams.

And that's only the beginning. Birds, squirrels, rabbits, dogs, cats, insects, fish, and many other creatures dine on many of the things contaminated by these chemicals—even the little patch of green in your backyard can be contributing to the problem.

Unfortunately, many of these chemicals cause health problems for humans, too, although there is considerable disagreement among government scientists about the cancer-causing potential of many chemical ingredients; some pesticides are rated as "probable" or "possible" carcinogens. Only a few ingredients have been evaluated for potential carcinogenicity by the Environmental Protection Agency, including **rotenone** (no consensus reached, but studies did produce tumors in rats), **propoxur** (rated as a "probable" carcinogen), and **benomyl** (a "possible" carcinogen). Some environmentalists and health officials take a stronger position on the cancer-causing properties of some other ingredients, including **benomyl, dicofil, glyphosate, malthion,** and **2,4-D**.

All of which leaves much of the guess work about the safety of pesticides up to you.

How to Read a Pesticide Label

Product labels are the Green Consumer's most important source of information about a pesticide's potential effect on the environment and on human health. Labels tell you how to mix and apply a product, which active ingredients it contains, and which pests it is intended to kill. *It is against federal law to use any pesticide in any way not in accordance with instructions on the label.*

The label below shows the information you will find on a typical pesticide label. The key to the highlighted numbers is on the facing page.

Back or Side Panel

1 STATEMENT OF PRACTICAL TREATMENT

IF SWALLOWED: Speed is imperative. Call a physician or Poison Control Center. Drink 1 or 2 glasses of water and induce vomiting by touching back of throat with finger. Do not induce vomiting or give anything by mouth to an unconscious person. Apply artificial respiration if breathing stops. IF ON SKIN: Wash thoroughly and immediately with cold running water and/or diluted vinegar. Do not use soap. IF IN EYES: Flush with plenty of water. Get medical attention if irritation persists.

2 PRECAUTIONARY STATEMENTS

HAZARDS TO HUMANS AND DOMESTIC ANIMALS. WARNING: May be fatal if swallowed, inhaled or absorbed through the skin. Do not breathe vapors or spray mist. Do not get in eyes, on skin, or on clothing. If spilled on clothing, remove and wash clothing before reuse. Keep away from children, domestic animals, and foodstuffs.

3 ENVIRONMENTAL HAZARDS

This product is toxic to birds and other wildlife. Keep out of lakes, streams, and ponds. Do not apply when weather conditions favor drift from treated areas.

4 DIRECTIONS FOR USE

It is a violation of Federal Law to use this product in a manner inconsistent with its labeling.

STORAGE: To be stored in original container and placed in areas inaccessible to children. PESTICIDE DISPOSAL: Product remaining in original container should be disposed of by securely wrapping trash. CONTAINER DISPOSAL: Do not reuse empty container. Rinse thoroughly before discarding in trash. NOTICE: Buyer assumes all risks of use, storage, or handling of this product not in strict accordance with directions given herewith.

DIRECTIONS FOR USE CONTAINED ON ENCLOSED INSTRUCTION SHEET.

5 **[Distributor name and address]**

Front Panel

NAME OF PRODUCT
Garden Spray

9

For use on ornamentals, fruits and vegetables. Controls: Aphids, Leafhoppers, Thrips, and similar sucking insects as specified on the instruction sheet.

8

ACTIVE INGREDIENT	
Nicotine expressed as alkaloid	40%
INERT INGREDIENTS	60%
TOTAL	100%

KEEP OUT OF REACH OF CHILDREN

WARNING **7**
See rear panel for statement
of practical treatment and
other precautionary statements

6
EPA Est. No. 5887-IL-1
EPA Reg. No. 5887-7
Net Contents 2 oz. Avdp.

Prepared by Cheryl Best and the staff of *Garbage* magazine.

Key to Pesticide Label

1. Directions for first aid in case of overexposure.

2. Cautions regarding use of the pesticide and its human health hazards.

3. A list of all known hazards to wildlife, beneficial insects, groundwater, etc., plus general instructions for avoiding these hazards.

4. Detailed instructions on how to mix and apply the pesticide and a list of all pests and crops for which the EPA permits its use. It is a federal offense to use any pesticide in any fashion not in accordance with these instructions.

5. Name of the company distributing the pesticide.

6. U.S. Environmental Protection Agency registration number (which signifies that the product has been registered with the EPA for the uses stated on the label) and establishment number (which indicates the facility where the pesticide was manufactured).

7. Signal word, listed in large letters, indicates toxicity: CAUTION indicates the product is relatively nontoxic or slightly toxic. WARNING means it is moderately toxic. DANGER/POISON indicates it is highly toxic.

8. Inert ingredients are substances in the formulation that do not kill the pests, listed by percentage only, not by name. Active ingredients are chemicals that kill the pests, listed by name as a percentage of the formulation.

9. The pesticide's common or brand name.

Gardening Without Pesticides

Do you really need pesticides—or at least as much as you've been using? Probably not. There is a growing awareness of alternative gardening methods, some using no chemicals at all, others using them selectively. Employing names such as "biological control," "integrated pest management," "conservation farming," and "genetic manipulation," they fill dozens of magazines and are the subjects of numerous books.

Here are some general guidelines for ways to control pests and weeds without using pesticides:

• Don't try to force plants to grow in environments that don't suit them. A rose that is grown against a house, for example, where it will be short of water, will inevitably suffer from mildew.

• If you know you have a particular pest or disease in your garden, try to find plants that are naturally resistant. Buy certified virus-free stock when planting soft fruit or potatoes.

• Don't plant large groups of one type of plant—plant a variety of plants. Gardeners have long believed, for example, that if you plant marigolds alongside carrots or potatoes, you can control pests like carrot fly or eel worm.

• Make sure you sow seed or plant seedlings at the right time. Vigorous growth is often a plant's best defense. Plant too early, and your plants will be unnecessarily vulnerable.

• Identify your specific problem. The more you know about the pest or disease that is damaging your plants, the less likely it is that you will use the wrong garden chemical. Often you will find that the problem will solve itself. The bean weevil, for example, eats notches out of broad bean leaves, but beans normally grow so fast that there is no danger to the crop.

• Pests such as caterpillars and sawfly larvae can often simply be picked off by hand. Diseased leaves, fruit, and other plant material should be cleared away. In this case, you may need to resort to a bonfire to ensure sterilization of material you may wish to return to the garden. The ash can be added to enhance your compost heap.

• Whatever you grow, encourage natural pest and disease controls. The best way to do this is to provide habitats and food plants for some of the insects and other creatures that prey on pests. A small pond could provide a home for toads and frogs, for example, whose diet includes slugs. Attract birds by providing nesting sites (bird boxes, hedges, and space behind climbers) and winter food (leave ornamental plants to go to seed and provide food dispensers). Grow hardy annuals like *Limnanthes douglassi* (the "poached egg" plant) and *Convolvulus tricolor*, which provide food for the syrphid fly, whose larvae consume large numbers of greenfly, blackfly, and other aphids. (See "Natural Remedies for Pests and Diseases" on page 182 for additional suggestions.)

• Protect crops with a physical barrier. If you are a Green Consumer, you probably don't buy plastic soda bottles, but one idea is to collect everyone else's plastic bottles and saw them in half. The ends can be used as mini-cloches for seedlings. Use old carpet padding around the base of your brassicas, which will help confuse the cabbage root fly.

• If, after all this, a particular pest gets out of hand and you feel you must use a chemical spray, use as little as possible.

If You Must Spray

• Read the product label before you buy. Check for particularly hazardous ingredients. Remember that most accidents with garden chemicals involve children under five years old; store all chemicals safely.

• Choose the least toxic pesticide—the ones with the signal word CAUTION on the label are considered less toxic than the ones labeled WARNING.

• Use only recommended dosages. Higher amounts rarely improve pest control, and can often lead to plant and soil damage.

• Never mix chemicals, unless the labels specifically tell you to do so. Wash out the mixer and sprayer before and after each spraying to avoid accidental mixing of chemicals. Be especially careful when changing from herbicides to insecticides or fungicides. Contaminating one with another may reduce their effectiveness.

• If you spray from too close a range, you can damage plants. Visible wetting of plant foliage with aerosol sprays is neither necessary nor desirable.

• Do not spray in bright sunshine. This can lead to plant damage, even if you are only spraying water.

• If you are growing vegetables or fruit, check the pesticide application instructions to ensure that you leave a long enough interval between spraying and harvesting. If the instructions on the label tell you to avoid touching or eating the plant or fruit for a period, chances are that the chemical will also be harmful to birds and other wildlife.

• Cover or remove exposed foods, fish tanks, and pet food and water dishes during and after application.

• Never spray near wells, streams, ponds, or marshes unless the instructions *specifically* allow for such use.

• Never apply to bare ground or eroded areas. When it rains, many pesticides bind tightly to soil and can be carried along with sediments to storm sewers and streams.

• Wash your hands immediately after applying the pesticide.

• *Never* dispose of surplus pesticides in or near ponds, streams, marshes, or ditches. Do not bury them in your yard, burn them, or dump them in a sewer or toilet. The best way to get rid of unused pesticides is to take them to a hazardous-waste site on a designated drop-off day. A growing number of cities and counties now offer such drop-off programs. If yours doesn't, urge them to do so by contacting city or county officials.

NATURAL REMEDIES FOR PESTS AND DISEASES

Insecticidal soap: This natural soap destroys pest membranes. It is effective against aphids, crickets, earwigs, mealybugs, rose slugs, scales, spittlebugs, white flies, and many others. One brand is **Safer's Insecticidal Soap.**

Bacillus thuringiensis: BT is a highly selective biologic insecticide that is particularly effective against leaf-eating caterpillars. It kills them by paralyzing the digestive tract. Brands include **Safer's Natural Caterpillar Killer, Dipel 2X, Javelin Liquid BT,** and **Entice Insect Feeding Stimulant.**

Milky spore: A natural bacteria that kills the grub phase of Japanese beetles. The milky spores actually remain alive in the soil, preventing new infestations for a few years.

Dormant oil sprays: Oil sprays, such as **Safer's Sunspray,** can be used to control scale insects, red spider, mites, mealybugs, and whitefly larvae on azaleas, evergreens, fruit trees, shade trees, shrubs, woody plants, and all other ornamentals.

Stale beer: Put this out at night to attract slugs.

Homemade sprays: One popular recipe calls for liquefying the following in a blender: 3 large onions, 1 whole garlic, 2 tablespoons hot red pepper in 1 quart of water. Stir in 1 tablespoon of soap. Apply with any spray bottle.

Insect-eating insects: These "good" bugs include ladybugs, lacewings, dark ground beetles, soldier beetles, and praying mantises, which will feast on some of your garden pests. These may be available in nurseries.

Natural Fertilizers. One form of protection against all sorts of pests and diseases is healthy soil. If you enrich it with the right fertilizers (as well as with humus and compost), your soil will be able to withstand insect attacks with greater success.

Before you fertilize, you may want to have your soil tested. You can

Storing Toxics

There are more cans and bottles of pesticides stored in Americans' homes than there are Americans, according to a 1992 study by the Environmental Protection Agency.

The National Home and Garden Pesticide Use Survey polled a sampling of U.S. households last fall to learn such things as the number and type of pesticides in our homes, how they are stored, and how they are disposed of. The survey looked at homes in 58 sample counties located in 29 states. All told, just over 2,000 households were surveyed. Among the findings:

- The number of pesticide products in the 2,078 residences was 7,945. From that, EPA estimates that the total number of products in storage nationwide is 324,538,000 (plus or minus about 22 million).
- Although pesticide labels warn consumers to store the chemicals securely—more than 4 feet off the floor and in a locked or childproofed room or cabinet—nearly half of households with children under 5 years old had improperly stored pesticides.
- Asked how they dispose of leftover pesticides, 13% said they took them to special collection sites and 67% said they put them with their regular trash. Most of the others used unspecified disposal methods. Six percent said they had not disposed of pesticides because they did not know how to do so safely.

Meanwhile, a useful publication called *Tackling Toxics in Everyday Products* offers to help us learn more about the real risks we face with pesticides and other products containing toxic ingredients.

The authors point out that although more than 17,000 chemicals are used in pesticides, cosmetics, food, and drugs, "less than 30% have been tested to the point where their ability to cause cancer or reproductive damage has been well defined." In addition, many of these chemicals contribute to environmental problems. In California alone, emissions from the use of consumer and commercial products account for about a fifth as much smog as created by cars, buses, and trucks.

The bulk of the 180-page book is a directory of organizations that have collected information on or are studying hazardous chemicals. The book's introductory material is also very useful, painting a portrait of the nature and extent of toxic chemicals in our daily lives. Tables include lists of products—from adhesives to air fresheners to all-purpose cleaners, listing the potentially harmful ingredients they may contain and whether they are likely to pose problems indoors, outdoors, or in waste disposal. The book is extensively cross-referenced, making it a comprehensive reference on the subject.

Copies of the book are available for $22.95 ppd. from Inform, 381 Park Ave. S., New York, NY 10016; 212-689-4040.

do this yourself by using soil test kits, or have a county extension agent analyze the soil for organic content and acidity. Acid soils can be balanced by applying powdered lime with a spreader.

Avoid artificial fertilizers in favor of organic manures. Artificial fertilizers can make the soil acid and drive worms away, and may also trigger rapid sappy growth in plants by releasing a burst of free nitrate. Ironically, too, overuse of fertilizers can damage overall soil fertility by "locking up" essential elements such as calcium and magnesium and by suppressing the activity of natural soil organisms that normally fix nitrogen from the air or make phosphate available to plants. In addition, excess fertilizers can reach your local stream and lead to water pollution problems.

Avoid applying fertilizer on windy days or just before a heavy rain. For best results, always apply fertilizers according to the directions on the package.

Mineral fertilizers should be used as a supplement to, rather than as a replacement for, natural nutrient recycling in the soil. In general:

Use

basic slag
bone meal
calcified seaweed
calcium sulfate
feldspar
fish meal
ground chalk
hoof and horn meals
limestone
magnesium
rock phosphate
rock potash
seaweed
unadulterated seaweed foliar sprays
wood ash

Restrict the use of

aluminum phosphate borax
dried blood
Epsom salts
hop waste
kieserite
leather meal
sulfate of potash
wood shoddy

Avoid all other mineral fertilizers, including

Chilean nitrate
muriate of potash
nitrochalk
quicklime
slaked lime
urea

The natural and organic garden suppliers listed on page 189 include many sources of worthwhile fertilizers.

HEALTHY LAWNS, HEALTHY PEOPLE

It is simply astounding how many chemicals Americans dump on their lawns in the never-ending quest for the immaculate, perfectly manicured, crabgrass-and-dandelion-free grass. Whether you apply these chemicals yourself or hire a gardener or lawn care company to do it, it's likely that the quest for the perfect lawn will inevitably lead you to apply one or more dangerous, unhealthy substances to your lawn.

Lawn care chemicals can make you—and your neighbors, your children, and your pets—sick. Most of the pesticides used in controlling weeds and insects are broad-spectrum biocides, which means they are poisonous to a wide variety of living organisms, including garden plants, wildlife, pets—and people. Inert ingredients, which may constitute 50 to 99 percent of a pesticide formula, may actually be more toxic than the active ingredients. The poisons can be absorbed through the skin, by the mouth, or by the breathing in of sprays, dusts, or vapors. You can be poisoned if you apply or are present during application; touch contaminated grass, shoes, clothing, or lawn furniture; or put contaminated objects (toys, golf balls) or fingers in your mouth. Moreover, the chemicals aren't necessarily safe once they dry. They can remain active for months, during which time they can release toxic vapors. Breathing these vapors, even from neighbors' lawns or while playing on or mowing contaminated grass, can make you sick.

The symptoms of lawn pesticide poisoning are deceptively simple and, unfortunately, similar to those of many other illnesses. Simply put, pesticides attack the central nervous system and other vital body centers. Symptoms include sore nose, tongue, or throat; burning skin or ears; skin rashes; excessive sweating or salivation; chest tightness; asthmalike wheezing attacks; coughing; muscle pain; headaches or eye pain; cramps; and diarrhea. Even harder to detect are some of the long-term problems, including lower male fertility, miscarriage, birth defects, and liver and kidney dysfunction.

Other frequent victims are birds, especially songbirds. What wildlife specialists have come to call the "lawn care syndrome" in birds refers to the classic signs of pesticide ingestion: shivering, excessive salivation, grand mal seizures, wild flapping, and sometimes screaming, according to one U.S. Fish and Wildlife Service wildlife toxicologist. While chemical-related deaths of bald eagles and other protected species have long been documented, songbird poisoning from lawn care products is a little-studied problem, one that began to surface only with the growth of gardening chemicals in the 1980s.

One of the biggest sources of lawn care chemicals are professional lawn care companies. According to *Lawn Care Industry*, a trade magazine, there are an estimated 5,000 companies that groom, mow, and spray more than 7 million lawns a year with an estimated 8 million pounds of pesticides—at a price tag to consumers of about $1.5 billion, nearly double that of a decade ago. But the actual price has been higher: Many of the chemicals used by these companies have posed a serious health threat. According to a study by Public Citizen of the 40 most commonly used lawn care pesticides, 9 were classified by the Environmental Protection Agency as "possible" or "probable" human carcinogens. Two of the herbicides, **chlorothalinol,** which sells under the trade name **Daconil,** and **diazinon** have been linked with the deaths of two golfers who died after playing on courses treated with these chemicals. (The government banned diazinon for use on golf courses in 1988, but says it does not have enough data on the long-term health effects of chlorothalinol to ban its use.) Public Citizen found one of the most widely used compounds in lawn care to be **2,4-D**, an ingredient in the defoliant Agent Orange. In 1986, the National Cancer Institute linked 2,4-D to an increased cancer risk among farmers who used it.

Despite this, many of the lawn care companies do little to educate their customers to the potential risks. Public Citizen surveyed company literature and found several companies saying just the opposite. For example, **Orkin Lawn Care** of Atlanta stated that its "fully approved and tested chemicals are completely child- and pet-safe." But says Public Citizen's pesticide specialist Laura Weiss, "To call any pesticide mixture 'safe' is inaccurate. Pesticides, by definition, are compounds that kill, and all chemicals used in lawn care service pose some risk to human health. The level of risk depends on a variety of factors, including the toxicity of the chemical, the extent of exposure, and a person's sensitivity to the chemical. Children are particularly vulnerable to increased cancer risks because their cells develop more rapidly than the cells of adults, and their play habits are likely to bring them into close contact with lawn care chemicals." (Copies of *Keep Off the Grass*, the Public Citizen study, are available for $10 for each of two editions, $15 for both, from Public Citizen's Congress Watch, 215 Pennsylvania Ave. SE, Washington, DC 20003; 202-546-4996.)

Not all lawn care services are bad. A growing number use natural or organic methods, and others will consider your concerns about pesticides in their treatment practices. **ChemLawn Services Corporation**, one of the national chains that formerly used 2,4-D (it stopped using the chemical in 1987), now offers a fertilizer-only package and is testing an organic lawn care program. For customers concerned about chemicals used in their lawn care, a ChemLawn spokesperson said it will "work with the customer" to find acceptable alternatives.

Solutions for Green Lawns. As with other aspects of gardening, a healthy lawn is its own best defense against weeds and pests. It will crowd out most weeds and resist insects and disease.

The three essentials of a healthy lawn, according to Stuart Franklin, author of *Building a Healthy Lawn* ($9.95 plus $2.75 for shipping from Storey Communications, Schoolhouse Rd., Pownal, VT 05261; 802-823-5811, 800-441-5700), are the roots, the top growth (the grass blades), and the soil. They all interact to give a lawn its level of health—or illness. The roots must go deep into the soil so they can seek out soil nutrients and find water during dry periods. Short mowing and light watering will prevent roots from going deep. Shallow-rooted lawns are the first to die or weaken during the summer heat. In this condition they attract insect and disease organisms.

Franklin, who owns and operates Naturalawn lawn care in Buffalo, New York, suggests the following steps to a healthy lawn:

In the Fall

- Fertilize. Grass will best use fertilizer in the fall.
- Rake leaves and add to a new or established compost pile for ready fertilizer in the spring.
- Service and store mowing equipment properly for trouble-free start-up in the spring.
- Purchase seed and limestone as needed for late winter application.

In the Spring

- Don't be overanxious to start gardening. Stay off the lawn if it is partly frozen or soggy.
- Gently rake off leaves and debris to prevent fungus disease and dead spots.
- Rake matted grass so it is standing straight up. Allow air and light

to penetrate.

• Make your first cuts short to clear off dead blades and encourage the grass to spread. Then raise your cutting height as the grass begins to grow vigorously.

• Don't cut off more than one-third of the blade at once; it's a shock to the plant.

• Seed bare spots and overseed thin spots by mid-spring. If grass isn't there, weeds surely will be.

• Fertilize early if you didn't do a late fall fertilizing. You want to get the grass tall and thick before weeds (especially crab grass) begin to sprout.

• Keep your blades *sharp*. Don't tear the grass. Cut it.

• Try using an organic fertilizer. Put some life back into your soil. Organic fertilizers and other organic supplements will also help to break up thatch.

• Leave short clippings on the lawn unless you have a thatch problem. Clippings won't normally cause thatch, but they will add to it once it is there.

Low-Impact Garden Supply Companies

With low-impact gardening (using minimal chemicals) or organic gardening (using no chemicals), the best place to start is with the seed. Several seed suppliers offer low- or no-chemical seeds. For example, at **Johnny's Selected Seeds** (Foss Hill Rd., Albion, ME 04910; 207-437-9294), seeds are grown with minimal pesticide use. **The Sprout House** (P.O. Box 1100, Great Barrington, MA 01230; 413-528-5200) offers organic, chemical-free sprouting seeds. For herb seeds raised with botanical pest control, contact **Meadowbrook Herb Garden** (Route 138, Wyoming, RI 02898; 401-539-7603).

Good examples of natural pest control products are the Safer brand manufactured by **Ringer Corporation** (9959 Valley View Rd., Minneapolis, MN 55344; 800-423-7544), including lawn, garden, and pet products made with nontoxic ingredients that biodegrade rapidly and leave no residues. Most Safer products fall into the EPA's "least hazardous" pesticide category and have been certified as biodegradable by Scientific Certification Systems. Safer products, available in stores and through many mail-order catalogs, include house plant products (such as insecticidal soap), garden products, soil and turf insecticides, herbicides, and insect traps.

Below is a list of select low-impact garden supply sources. Many offer a comprehensive range of lawn and garden care products; others do not. Be sure to inquire for the products you are interested in before asking for a catalog.

Gardens Alive!
5100 Schenley Pl.
Lawrenceburg, IN 47025
812-537-8650

Growing Naturally
P.O. Box 54
Pineville, PA 18946
215-598-7025

Green Pro
380 S. Franklin St.
Hempstead, NY 11550
516-538-6444, 800-645-6464

Integrated Fertility Management
333 Ohme Gardens Rd.
Wenatchee, WA 98801
509-662-3179, 800-332-3179

Meadowbrook Herb Garden
Route 138
Wyoming, RI 02898
401-539-7603

Mellinger's
2310 W. South Range Rd.
North Lima, OH 44452
216-549-9861, 800-321-7444

Nature's Control
P.O. Box 35
Medford, OR 97501
503-899-8318

Nitron Industries, Inc.
4605 Johnson Rd.
P.O. Box 1447
Fayetteville, AR 72702
501-750-1777

The Necessary Trading Co.
One Nature's Way
New Castle, VA 24127
703-864-5103, 800-447-5354

Ohio Earth Food, Inc.
5488 Swamp St. NE
Hartville, OH 44632
216-877-9356

Ringer
9959 Valley View Rd.
Eden Prairie, MN 55344
612-941-4180, 800-654-1047

Universal Diatomes, Inc.
3434B Vassar St. NE
Albuquerque, NM 87107
505-247-3999

Pheromones, the natural sex-attractants of insects, are being applied to traps and used as pest control devices. Two companies offering pheromone trapping systems are:

Consep Membranes Inc.
213 SW Columbia St.
Bend, OR 97702
503-388-3688

Insects Limited, Inc.
10540 Jessup Blvd.
Indianapolis, IN 46280
317-846-3399, 800-992-1991

COMPOSTING

Composting is the process by which organic wastes—including food wastes, paper, and yard wastes—can decompose naturally, resulting in a product rich in minerals. Gardeners love compost—it makes an ideal soil conditioner, mulch, resurfacing material, or landfill cover—and so do plants.

During composting, a mass of biodegradable waste, combined with sufficient moisture and oxygen, "self-heat"—a process by which microorganisms metabolize into organic matter and release energy in the form of heat. The process is nothing more than an accelerated version of the breakdown of organic matter that occurs under natural conditions, such as on the forest floor. Because composting is a natural process, it can be carried out with as little, or as much, intervention and attention as the composter desires.

While several million Americans have ongoing compost piles in their yards, the growing action is going on in communities, which are finding that large-scale composting of leaves and other yard waste not only reduces solid waste but actually can create a marketable—and profitable—final product. About 1,000 cities and counties now have leaf-composting facilities, according to the Environmental Protection Agency. Communities variously give the compost away or sell it to residents, use it for public park service projects, sell it to other communities, or trade it for nursery stock. With yard wastes constituting about a fifth of all household waste, this material is a continuously renewable resource.

Examples abound of communities turning their garden trash into cash. For more than a quarter century, for example, the borough of Swarthmore, Pennsylvania, has collected about 400 truckloads of leaves annually, which they spread over a vacant lot owned by Swarthmore College. Each August, the leaves are bulldozed into a large pile, adjacent to piles from three prior years. Allowed to decompose over a two- to three-year period, the borough is left with nearly 100 loads of rich leaf mold in the oldest pile. That, in turn, is sold to residents for $25 per truckload delivered.

Starting your own backyard compost heap is easy. Simply gather fallen leaves, dead plants, and brush in a corner of your backyard. Bacteria, fungi, and other organisms will break it down, eventually

How to Compost Indoors

You don't need a large outdoor space—or even a backyard—in order to compost your kitchen scraps.

Making compost indoors is like making it outdoors but on a smaller scale, says David Goldbeck, author of *The Smart Kitchen* (Ceres Press, 1990).

Wastes should be chopped or pureed in the blender or food processor, or you can use only foods that are already quite small, such as coffee grounds, tea leaves, vegetable parings, and the like. Use any kind of container. The wastes should be layered with soil (make sure it is not sterilized), and after the first few days stir it *daily*. Small amounts of grass clippings (from a suburban friend) or hay (from a stable) held in a plastic bag can also be mixed in. It is the mixing and stirring that will kill off any odor-causing bacteria. Keep it moist but not soggy, and do not add garbage to a container that is already in the process of composting; start a new one.

The decomposition should take about two weeks, yielding fertilizer or humus for houseplants, windowbox, greenhouse, or an unusual gift for a plant-loving friend.

A word of warning: If you are mold sensitive, you should probably be cautious about storing food wastes.

From design and construction to operation and cleaning, Goldbeck's well-written book offers a wide range of other useful information that can turn any kitchen into a "greener" and healthier one.

The Smart Kitchen is available for $17.95 postpaid from Ceres Press, P.O. Box 87, Woodstock, NY 12498.

creating compost, a loose, crumbly earth. The key is to achieve a proper balance of moisture and oxygen. Experts say to keep the pile under five feet tall and to keep it moist, but not wet. Layer the pile using green leaves (which supply oxygen) and brown leaves or straw (which supply carbon).

Whether you compost or not, don't collect yard wastes in plastic bags. A better alternative is the **Ecolobag** ($30 for 50 bags from **Dano Enterprises, Inc.,** 75 Commercial St., Plainview, NY 11803; 516-349-7300), a heavy-duty, weather-resistant 50-pound kraft paper bag specifically designed for collecting leaves and other biodegradable trash. The convenient bag has a 12-by-16-inch square bottom that allows it to stand by itself.

HOW TREES SAVE THE EARTH

You may think of trees as a gift from Mother Nature, something to climb up or sit under, or the cause of fall raking. You probably don't think of trees as life savers. But that's exactly what they are. In cities, countryside, and forests, trees breathe life into our planet and save it from a host of environmental problems.

What exactly do trees do? Aside from their beauty and the food some of them produce, some of their other gifts include:

- **Cooling cities.** Urban areas are "heat islands"—their buildings, streets, cars, and other structures and activities soak up heat on a summer's day and release it at night. Researchers at the Lawrence Berkeley Laboratory in California found that average temperatures in the city can be five to nine degrees higher than those in surrounding suburbs. Groups of trees can offset this heat, operating as nature's air conditioners. Trees also help to reduce noise in cities.
- **Battling the greenhouse effect.** Because they absorb carbon dioxide, trees and other greenery offer the cheapest way to combat the greenhouse effect. (Carbon dioxide is responsible for about half of the greenhouse problem.) The average tree absorbs between 26 and 48 pounds of carbon dioxide a year, according to the American Forestry Association. An acre of trees takes in about 2.6 tons of carbon dioxide, enough to offset the emissions produced by a car driving 26,000 miles.
- **Preventing erosion.** Trees protect against the erosive power of wind, helping to protect topsoil and retain soil moisture. Deprived of their protective tree cover, hillsides are easily eroded and less able to retain rainfall. Without trees to break its force, the wind finds the exposed topsoil easy pickings. Continued wind produces giant, gritty clouds that steadily diminish America's precious soil heritage.
- **Reducing energy needs.** The shade provided by trees can save considerable energy and money. In the summer, three well-placed trees around a house can cut home air conditioning energy needs by 10 to 15 percent.

With these benefits in mind, several local and national organizations have created ambitious campaigns to plant millions of new trees across the United States as well as in other parts of the world:

Cutting Down on Cutting Down

This holiday season, another 50 million evergreens will fall to the woodcutter's ax (or, more likely, chainsaw). Worldwide, the number is around 80 million trees. Many of these come from specially harvested tree farms, so cutting them down doesn't contribute to environmental problems. (But disposing of the trees does; more on that in a minute.)

The New Green Christmas, a handy guide from The Evergreen Alliance published by Halo Books, offers some welcome insight into how to cope with tree cutting, trimming, and disposal during this season. Here are some tips:

• Consider growing your own. It needn't be an evergreen. Just about any large plant or small tree can serve as a place to hang your ornaments. You don't need a plot of land to do it; you can buy or grow a large potted plant.

• Don't plan to keep a living tree indoors more than 2 weeks, 3 weeks maximum. And don't position it close to heat sources, including an unshaded south-facing window.

• Think carefully about where you want your tree to live until next year's holiday: in its container or in the ground. Consult your nursery for advice.

• If you use a cut tree, try to keep it from being dumped in a landfill. (18% of landfills consists of yard waste.) One option is to take it to a local nursery that has facilities for composting or mulching trees. Even better: Add it to your own compost pile. If it's a pine tree, save the needles; they make great mulch for your garden.

• Although burning dry branches and brittle needles makes a good, crackling fire, it's not a good practice. It distributes pine tar in your flu and chimney.

• What about non-living trees? If you must buy one, make sure it gets the longest use possible. When you no longer want it, don't throw it away. Donate it to a charitable organization that can use it.

• **The Basic Foundation** (P.O. Box 47012, Saint Petersburg, FL 33743; 813-526-9562), has created a campaign to plant endangered tropical trees in Costa Rica and Nicaragua. Tree planting takes place in community nurseries of small farmers that are trained in the reproduction, planting, and care of endangered native timber trees as well as high-quality fruit trees. In return for the seedlings and technical assistance, the farmers reforest degraded watersheds and areas not suitable for other agriculture. The Basic Foundation accepts tax-deductible contributions: $5 pays for a single tree; $250 for planting a hectare—approximately 1,000 trees.

• **TreePeople** (12601 Mulholland Dr., Beverly Hills, CA 90210; 818-753-4600) is a nonprofit group whose mission is to help heal the environment by challenging citizens of the Los Angeles area to participate in the planting, care, and appreciation of trees and the urban forest. Programs include Citizen Forester training for individuals who want to organize a planting in their neighborhood, and Campus Forestry for high school students who want to plant at their schools. The organization's book, *The Simple Act of Planting a Tree*, based on the Citizen Forester program, explains how to organize a neighborhood project. The book is available in many bookstores.

• **Trees for Life** (1103 Jefferson, Wichita, KS 67203; 316-263-7294) is responsible for having planted 14 million fruit trees in India, Brazil, Guatemala, and Nepal. The group also is trying to organize a program in which the United States and the former Soviet Union would help to plant 100 million trees in underdeveloped countries. The group seeks donations from individuals and companies.

• The **American Forestry Association** (1516 P St. NW, Washington, DC 20005; 202-667-3300) has created a program called "Global ReLeaf," whose aim is to plant 100 million new trees by 1992 to help reverse the greenhouse effect and slow global warming. Several cities have made significant commitments to the campaign; Los Angeles alone has set a planting goal of 5 million trees. According to AFA, there are about 100 million "planting spaces" in the United States where trees could shade homes and businesses. According to Lawrence Berkeley Laboratory, planting those 100 million trees could save American consumers an annual *$4 billion* in energy costs alone. Contact AFA to find out if there is a Global ReLeaf program in your area, or how to set one up if there is not.

• **The National Arbor Day Foundation** (100 Arbor Ave., Nebraska City, NE 68410; 402-474-5655) has a campaign to promote tree planting in the United States. The nonprofit organization offers ten Colorado blue spruce trees (six to twelve inches tall), with planting instructions and a copy of *The Tree Book*, with each $10 membership contribution. (People in southern and West Coast states receive ten bald cypress trees, conifers better suited to those climates.)

DEALING WITH PETS' PESTS

In terms of size, they're, well, fleas and ticks. But these tiny creatures can wreak large-scale havoc. On pets, they can spread disease and parasitic infections such as tapeworms and Lyme disease. Bored with frolicking in fur, they may decide to feast on humans, causing no end of torment. And like other bugs, once they've ingratiated themselves in your cozy home, they have little intention of leaving.

So what do you do when these tiny, thick-skinned, blood-sucking insects, these stealth bombers of the bug world, hitch a ride into your home on the back of your dog or cat?

For many people, particularly those prone to get bitten, no means of destruction is too kind. Sprays, powders, shampoos, pesticide bombs—whatever works is fair game. But in this bloodthirsty quest to eradicate these bloodthirsty bugs, there's an environmental price to be paid. In the end, we may be doing more damage than we realize to ourselves, our homes, and our environment—not to mention our pets.

Ignominious Ingredients

Understanding the problems with pet pest products begins with reading a few product labels. As you do, you'll find most commercial flea and tick sprays contain carbaryl, a pesticide known by its trade name, Sevin. It works by attacking pests' nervous systems.

"Carbaryl is really nasty," is how Susan Cooper puts it. Cooper, staff ecologist with the National Coalition Against the Misuse of Pesticides (NCAMP), says carbaryl is a teratogen, meaning it can harm developing fetuses in both animals and humans. The Environmental Protection Agency has warned that products containing carbaryl should not be used on pregnant animals, although they have not similarly warned pregnant women against administering such products.

"I am appalled that the EPA allows this compound to be used on household pets—animals that may be petted and touched by small children," says Cooper. "There is a growing body of evidence of wide neurological effects." Carbaryl appears to affect cats more than dogs, she says, causing moodiness and erratic behavior in normally docile animals, anything from attacking humans to killing smaller animals.

Carbaryl "is one of the more dangerous pesticides around," says Phillip Dickey of the Seattle-based Washington Toxics Coalition. In addition to whatever harm it may do to pets and humans, carbaryl is highly toxic to bees, he says. That, in turn, can result in other serious ecological problems because many flowering plants rely on bees for pollination.

The EPA, which is in the process of "re-registering" carbaryl, sees no conclusive evidence of serious problems. "In 24 studies over 20 years, there have been some negative reports, some positive, and some marginal," says Dennis Edwards, an EPA pesticides program official. "There is no indication of a problem if used properly."

Perhaps. But what exactly is "proper use"? According to Virginia veterinarian Nina Miller, pet owners often go "flea crazy," figuring that if a little pesticide is good, a lot is better. The unfortunate result may be pet poisoning.

Most poisoning cases Miller sees relate to a category of pesticides called organophosphates, such as dichlorovos, another common ingredient in flea and tick powders and sprays. "I don't like using them, but sometimes the infestation is so bad that they're necessary," she says. "There is usually no problem if they're administered correctly."

But in some cases, even proper use of these products results in disaster. Dickey tells of a "rash of incidents" related to dichlorvos in which "cats were dying just from wearing the collars."

Another problem has to do specifically with commercial foggers, products designed to "bomb" one or more rooms with pest-killing ingredients. Many such products rely on petroleum distillates as propellants—the ingredients that force the product out of the can when you press the button. These pose a fire hazard when used in close contact with, say, a burning pilot light (on stoves or water heaters, for example), or even hot electric motors (such as the one in your refrigerator); the distillates can ignite fires or even explosions. Cooper advises that when using foggers you should turn off all motors and pilot lights.

There is one product about which most environmental experts are particularly wary: Blockade, a flea and tick spray from Hartz Mountain. The spray is a combination of two compounds not usually known to cause problems: DEET, a widely used insect repellent, and fenvalerate, one of a family of insecticides known for their relatively low toxicity.

But in 1987, after complaints by pet owners of violent reactions to the sprays by their pets—from salivating and diarrhea to tremors and death—Hartz voluntarily recalled the product. A month later, the

Birds Doo It, Bees Doo It

We've recently become enchanted with Zoo Doo, a "premium organic fertilizer" made from animal waste from local zoos. Zoo Doo Compost Company has arranged to collect droppings from elephants, rhinos, giraffes, and the like. After the manure has been composted for 6 months, it is dried, screened, then blended with a variety of ingredients. Just over half of the profits from sales of Zoo Doo—a 15-pound pail goes for $13—are donated to the participating zoos. An environmental bonus: the animal wastes are kept out of landfills. The product was introduced at the Memphis Zoo, where it has enjoyed the sweet smell of success. Zoo Doo president Pierce Ledbetter says he is presently working with several zoos across the country "and hope that each will have its own Zoo Doo in time for spring." For more information, call 800-I-Luv-Doo.

company returned the same product to shelves with additional labeling, warning, among other things, that the spray shouldn't be used on dogs younger than three months, cats under a year, or on pregnant, sick, or old animals.

But Cooper and others believe that some animals not fitting this profile may also be at risk. Also problematic is that symptoms do not always show up immediately, making it difficult at times to link them with use of Blockade.

Moreover, according to George LaRocca, who has followed the case at the EPA, there is another potential problem: besides killing insects, fenvalerate is also toxic to aquatic life. Everything from pets' bath water runoff to improper disposal of used product containers could cause problems. Says LaRocca: "We are concerned about how it's used."

Taming Toxicity

Fortunately, there are less toxic alternatives to Blockade and other problematic products. For example, there is a new breed of pesticides containing pyrethrin, a natural ingredient derived from the plant *Chrysanthemum cinerariaefolium*, an African daisy. Pyrethrin works on insects by increasing their neurological activity, thereby overstimulating them; in effect, the insects burn themselves out. Synergistic chemicals are usually added to pyrethrin formulations to increase their effectiveness. One of the safest is piperonyl butoxide, derived from

Brazilian sassafras. Pyrethrin products usually can be used both directly on pets as well as around the house.

Another benefit of pyrethrin is that it degrades relatively effectively, resulting in little or no water contamination. However, only one such product has been certified as biodegradable by Green Cross Certification Company: the Safer line of products made by the Ringer Corp., based in Minneapolis.

"Our goal has always been to make a product that was safe for pets and humans as well as the environment," says Rob Ringer, the company's director of public relations. Besides its flea and tick killer, the Safer line also includes insecticidal soaps made from potassium salts of fatty acids, which are no more toxic than ordinary hand soap.

Still, as with all pesticides, even pyrethrins must be used with care. Experts warn that people and animals with asthma, particularly those allergic to ragweed pollen, may react to pyrethrin. And pyrethrin also can be toxic to aquatic life and should not be used near water.

Baby Killers

All the pesticides we've discussed so far assume that you're dealing with full-grown adult pests. After all, they're the ones that do the most damage, including laying eggs in carpeting, cat litter, upholstery, and anywhere else they can find organic matter.

But the chemicals that kill adults won't generally harm eggs, pupae, and larvae. To get rid of them, too, requires another category of pesticide called a "growth regulator." Growth regulators interrupt the development of sub-adults by rendering the larva incapable of developing from a pupa to an adult.

"I don't think you have a good flea control program without growth regulators," says the EPA's Edwards. "They're extremely safe compounds. I've heard of no complaints about them. One brand is even certified for use against mosquitoes in drinking water." However, according to NCAMP, laboratory experiments have shown that pests can develop an immunity to the growth regulators after prolonged exposure.

Still, growth regulators are the best thing going to get rid of both present and future infestations. The most effective growth regulator is called methroprene, which can be found in Precor, made by Zoecon.

In the end, dealing with pests in a nontoxic way takes a bit more effort. But even harsher chemicals must be applied over and over. Over

Don't Poison Your Pets

Gardens are one of the favorite foods of snails during the spring. And snail poison is one of the favorite foods of household pets during that same period. The sweet flavor and foodlike appearance of snail bait is attractive to pets. But the bait can be as poisonous to dogs and cats (and most other small animals) as it is to snails. As an alternative to snail bait, you might try putting a wire mesh fence made of hardware cloth around the garden; hardware cloth is available at most hardware stores. Plant it so it sticks up a few inches above the ground, then curve the top out, away from the garden. It is difficult even for a snail to walk on wire upside down.

Snail bait isn't the only garden hazard. Mole and gopher killers contain strychnine, a sure animal killer. Planting pop bottles at gopher hole entrances (the animals don't like the sound of the wind whistling over the bottle tops) or placing toy windmills in the same area (the vibrations that the windmills transmit to the ground disturb the pests) is a safer and more interesting way to encourage gophers and moles to move on. Chemical herbicides used on dandelions and other weeds can prove troublesome, too, should your pet feast on the treated grass—or even lick its paws after playing in it. To protect your pet's health, always store insecticide sprays and powders out of harm's way, and use them with caution.

If you suspect your pet has accidentally ingested yard and garden poisons, symptoms such as vomiting, trembling, convulsions, dilated pupils, salivation, labored breathing, weakness, or collapse will generally follow. If this happens, try to determine what the poison was, and when and how much was ingested—and call your veterinarian immediately.

By the way, animal pesticides are harmful to humans, too. A few years ago, illnesses among workers at a Georgia veterinary hospital prompted federal health officials to warn against excessive exposure to pesticides, particularly fenthion, the one the Georgia clinic used to kill fleas. Four workers at the clinic complained of symptoms, including shooting pain, muscle weakness, and numbness. Chronic exposure to pesticides such as fenthion has been shown to cause nerve damage, according to the Centers for Disease Control.

the long run, minimizing poisons will be worth the effort—for you, your pets, and everyone else.

Winning the War on Bugs

Perhaps the most important question to ask when seeking to control pets' pests is whether pesticides really are necessary at all. "Our general feeling about the problem is that people are using pesticides as a first resort and only tool," says Phillip Dickey. "That is not the right way to go about it. There are a variety of tools available."

Dickey advocates using relatively harmless substances, such as applying boric acid powder deep within carpeting, which acts as a dessicant: it makes the environment within the carpet too dry for the insects to survive. Boric acid is available in most hardware stores. It is best applied by a professional carpet cleaner, to ensure that the ingredients reach far enough into the carpet to do their job; it needs to be almost literally pounded into the carpet. Dessicants probably are not wise to use around small children, who might absorb the ingredients while crawling on the carpet. A safer alternative might be steam-cleaning the carpet.

Cleaning, in fact, is key. For bad infestations, daily vacuuming can help. It's important to change the vacuum bag immediately after vacuuming.

Cleaning the pet is important, too. Daily cleaning of a dog or cat with a fine-toothed flea comb, available at any pet store, removes existing fleas. (You must then flick the fleas into soapy water to kill them.)

Don't try to manage fleas outdoors. They're everywhere and you can't get rid of them. Resign yourself to the fact that if your pet goes outdoors, you're going to have to deal with fleas indoors.

What about homeopathic and other alternative remedies? There are a variety of herbs that are said to render pets unappetizing to insects. But we weren't able to find any that were heartily endorsed. And some are dubious at best. For example, there's Dr. Goodpet, from Very Healthy, Inc., in Inglewood, CA. In addition to an abundance of herbs, the product contains 20 percent alcohol. The pet is supposed to drink this stuff. Whether or not it kills the pests, it's probably not too good for Fido or Fluffy.

What about those high-tech electronic flea collars advertised in magazines as environmentally preferable? Forget about them. As the EPA's Dennis Edwards puts it, "We'd be delighted if one did work, but as yet they don't."

Souces for Alternative Pest-Control Products

Many of the companies listed below offer free catalogs and product information sheets.

Anti-Litter Campaign

If you are among those who have pondered the contribution of kitty litter to the nation's trash heap, you'll be glad to know about Yesterday's News, a brand of cat litter made from recycled newspaper. The product's distributor claims its product to be more absorbent and less smelly than traditional clay litter. Moreover, the used litter is said to make an excellent soil conditioner, compost, or mulch. And we haven't even mentioned the potential to keep newspapers out of landfills. A 5-pound bag of Yesterday's News sell for about $2. To order, contact Yesterday's News, P.O. Box 182, Swampscott, MA 01907; 617-592-3373.

INSECT GROWTH REGULATORS

• **Zoecon** (12005 Ford Rd., Dallas, TX 75234; 800-527-0512) manufacturers Precor products.

PYRETHRIN- AND NATURAL-BASED PRODUCTS

• **All the Best Pet Care** (2713 E. Madison Ave., Seattle, WA 98112; 800-962-8266) carries natural flea-control remedies and other products.

• **Eco-Safe Products, Inc.** (P.O. Box 1177, St. Augustine, FL 32084; 800-274-7387) makes natural pyrethrin-based flea powder.

• **Frontier Cooperative Herbs** (Box 299, Norway, IA 52318; 800-669-3275) sells a variety of natural pet products.

• **Morrills' New Directions** (P.O. Box 30, Orient, ME 04471; 800-368-5057) carries a wide range of alternative pet products. Write or call for a free copy of *Your Natural Pet Care Catalogue*.

• **Necessary Trading Co.** (One Nature's Way, Newcastle, VA 24127; 703-864-5103) is a supplier of organic flea and pest-control products.

• **Pet Guard** (P.O. Box 728, Orange Park, FL 32073; 800-874-3221) sells a variety of natural pet products.

• **Pristine Products** (2311 E. Indian School Rd., Phoenix, AZ 85016; 602-955-7031) sells organic flea and tick control products, combining food-grade diatomaceous earth and natural pyrethrins.

• **Ringer Corp.** (9959 Valley View Rd., Minneapolis, MN 55344; 800-423-7544) makes Safer brand pest-control products, available at many hardware and garden-supply stores.

Flea Combs

• **Breeders Equipment Co.** (Box 177, Flourtown, PA 19031; 215-233-0799).

• **Petco Animal Supplies** (9151 Rehco Rd., San Diego, CA 92121; 800-765-9878).

HOME ENERGY & FURNISHINGS

There is no place like home when it comes to being a Green Consumer. Aside from the food we bring into it, the trash we take out of it, the gardens we plant around (or inside) it, and the vehicles in which we drive to and from it, there is the matter of the home itself—in particular, the energy needed to keep its occupants cozy and comfortable.

The cost of that energy is significant, by any measure. Homes are the third largest users of energy (after industry and transportation), representing about 20 percent of all energy consumed in the United States, at a cost of about $130 billion, just under 3 percent of our nation's gross national product. Producing that energy—through the thousands of coal-fired, nuclear, hydroelectric, and other power-generating facilities—costs the earth dearly in terms of the resources it requires and the pollution it creates. Electrical utilities are primary contributors to acid rain (through the sulfur dioxide emissions of coal-fired power plants) and greenhouse warming (through the burning of fossil fuels such as coal, oil, and natural gas). (See "Seven Environmental Problems You Can Do Something About" for more on this.)

Worst of all, such problems needn't exist. Estimates of the amount of energy wasted in U.S. homes through inefficiency range from 40 to 70 percent. Conservatively, most experts agree that we could easily save about half of our home energy costs—along with similar cuts in pollutants—if we used our energy resources wisely, including purchasing energy-efficient appliances. That wasted energy is also wasted money—hundreds of dollars a year for some households. As you'll see, the purchase price of most appliances pales compared to the cost of operating them over their lifetimes.

Wasted energy isn't the only item about which Green Consumers should be concerned at home. Water use is another big concern, particularly in drought-prone sections of the United States. Those areas used to be limited to western states, but in recent years the Midwest and the populous portions of the East Coast also have faced water-shortage problems, with most forecasts indicating even greater problems still to come.

There's more. Many home furnishings, from carpets to couches to cabinets, are made from materials that create additional pollutants either in their manufacturing process or when disposed of, or which come from threatened rain forest timber. If you understand the potential problems, it is not difficult to choose products that don't contain these environmentally unfriendly components.

In short, there's plenty that Green Consumers can do right in their own homes, with little inconvenience or disruption to their lifestyles.

ENERGY-EFFICIENT APPLIANCES

Equipping your home with electrical and gas appliances that use less energy is more than environmentally sound—it's downright profitable. Appliances and heating and cooling equipment cost an average American household more than $1,000 per year. You can sharply cut those costs—and help reduce the need for building polluting and resource-draining power plants by using high-efficiency appliances.

You may be surprised to learn which appliances in your home guzzle the most electricity. You may have been taught always to turn off lights when you leave the room, for example, but lights are the least of the problem. Your refrigerator, quietly (or not-so-quietly) gurgling in the corner of the kitchen has a far bigger electrical appetite. In fact, cumulatively, America's refrigerators use the output of about 25 large power plants—about 7 percent of the nation's total electricity consumption and more than half of the power generated by the nation's nuclear power plants. According to one estimate, if every

The Ten Key Energy Users.

Not including your central heating system, here are the ten things likely to be the biggest energy users in your home:

1. Water heaters
2. Refrigerator/freezers
3. Freezers
4. Air conditioners
5. Ranges
6. Clothes washers
7. Clothes dryers
8. Dishwashers
9. Portable space heaters
10. Lighting

household in the United States had the most energy-efficient refrigerators currently on the market, the electricity savings would eliminate the need for about ten large power plants.

Federal law now establishes minimum efficiency standards for major home appliances and heating and cooling equipment. The standards will require all new appliances to be 10 to 30 percent more efficient than the average models sold previously. It has been estimated that between now and the year 2000, these standards will save consumers at least $28 billion in energy costs over the lifetimes of the products—about $300 per household. Moreover, the federal standards will reduce peak electricity demand by the equivalent of 25 large power plants.

There's no question that energy-efficient appliances cost more to purchase than inefficient ones, but that scarcely matters: *the purchase price is by far the smallest cost of owning an appliance*. For example, the cost of energy to run a refrigerator for 15 to 20 years typically is three times as much as the refrigerator's purchase price. (See the box "Determining Life Cycle Costs.") So buying efficient appliances is a good long-term investment. And as energy costs continue to rise—the electricity rates have more than tripled during the past 15 years and are certain to continue going up—the investment in energy-efficient appliances will yield even bigger dividends.

Comparing Appliances

How can you tell which appliances are the most efficient? Seven major appliances—boilers, clothes washers, dishwashers, heat pumps, refrigerators and refrigerator/freezers, room air conditioners, and water heaters—are required by law to disclose their energy efficiency ratings (EERs) or annual energy costs on yellow Energy Guide labels attached to new appliances. These costs and ratings are derived from standardized industry tests conducted by manufacturers. Other appliances are more difficult to compare, although there are energy-saving features worth looking for.

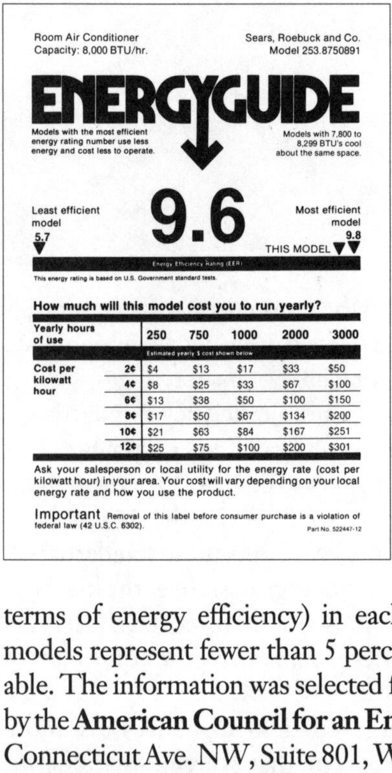
Room Air Conditioner
Capacity: 8,000 BTU/hr.

Sears, Roebuck and Co.
Model 253.8750891

ENERGYGUIDE

Models with the most efficient energy rating number use less energy and cost less to operate.

Models with 7,800 to 8,299 BTU's cool about the same space.

Least efficient model 5.7

9.6

Most efficient model 9.8

THIS MODEL

Energy Efficiency Rating (EER)

This energy rating is based on U.S. Government standard tests.

How much will this model cost you to run yearly?

Yearly hours of use		250	750	1000	2000	3000
		Estimated yearly $ cost shown below				
Cost per kilowatt hour	2¢	$4	$13	$17	$33	$50
	4¢	$8	$25	$33	$67	$100
	6¢	$13	$38	$50	$100	$150
	8¢	$17	$50	$67	$134	$200
	10¢	$21	$63	$84	$167	$251
	12¢	$25	$75	$100	$200	$301

Ask your salesperson or local utility for the energy rate (cost per kilowatt hour) in your area. Your cost will vary depending on your local energy rate and how you use the product.

Important Removal of this label before consumer purchase is a violation of federal law (42 U.S.C. 6302).

Part No. 522447-12

Beginning on page 208 are listings of the most energy-efficient appliances in ten categories: refrigerators, freezers, clothes washers, dishwashers, gas furnaces, oil furnaces, water heaters, room air conditioners, central air conditioners, and central heat pumps. Due to space limitations, only the five top-rated models (in terms of energy efficiency) in each category are included. These models represent fewer than 5 percent of all models currently available. The information was selected from a much larger list, compiled by the **American Council for an Energy-Efficient Economy** (1001 Connecticut Ave. NW, Suite 801, Washington, DC 20036; 202-429-8873), a nonprofit research group, and is reprinted here with permission from ACEEE. To obtain the *Consumer Guide to Home Energy Savings*, the most complete and up-to-date list of appliance energy ratings, send $8.95 to ACEEE, 2140 Shattuck Ave., Ste. 202, Berkeley, CA 94704.

Keep in mind that energy performance is only one of several important criteria for selecting home appliances. These ratings, for example, do not consider product reliability and other features such as

Determining Life-Cycle Costs

The real cost of an appliance can't be judged when it is purchased, or even over the course of a single year. The most accurate measure is the "life-cycle cost"—the cost to own and operate an appliance over its lifetime. In short, the life cycle cost is:

Purchase Price + Annual Energy Cost x Estimated Lifetime = Life-cycle Cost

For example, if you purchase a refrigerator for $650, it costs about $150 a year to run. If it lasts for 20 years, the life-cycle cost would be

$650 + ($150 x 20) = $3,650

To help you determine life-cycle costs, here are the average life spans of major appliances:

Appliance	**Average life span (years)**
Air conditioner (central)	15
Air conditioner (room)	12
Clothes dryer	18
Clothes washer	13
Dishwasher	12
Freezer	20
Range/oven	18
Refrigerator/freezer	20
Water heater	13

Compare, for example, Refrigerator A, which costs $500 and costs $125 a year in energy, with Refrigerator B, which costs $750 and costs $105 a year in energy. Life-cycle costs would be as follows:

Refrigerator A = $500 + ($125 x 20) = $3,000
Refrigerator B = $750 + ($105 x 20) = $2,850

So, even though Refrigerator B costs $250 more to buy, the life-cycle costs will actually be $150 less than those for Refrigerator A over the life cycle of the appliance.

freezer capacity, access to shelves, and options such as ice makers and specialized storage bins. However, energy-efficient appliances are usually high-quality products due to the advanced materials and manufacturing processes used in their construction. To compare other appliance features, consult *Consumer Reports*, which regularly provides in-depth analyses of major appliances.

An asterisk (*) appearing in a model number indicates a digit or letter that varies with the color or style of the appliance.

Refrigerators

Today's energy-efficient refrigerators use about one-third less electricity than comparable models ten years old or older. To gain the maximum efficiency from a refrigerator, it's important to shop for the size and features that best suit your needs. If you buy one that's too big, you'll waste energy. A refrigerator that's too small won't cool foods as efficiently. A rule of thumb is that one to two people need a model with an interior capacity of at least 12 cubic feet. Three to four people need 14 to 16 cubic feet. For more than four people, add two cubic feet for each additional person.

The models on the following pages are grouped by door style, defrosting capability, and volume (size), and are listed in order of increasing energy use. If two different-size refrigerators use the same amount of electricity per year, the larger model would be considered more efficient, because it keeps a greater amount of space cold with the same amount of energy.

The annual energy costs shown are based on an electricity price of 6.75 cents per kilowatt/hour (kwh), which is the electricity price used on the Energy Guide labels. Your actual cost may differ depending on the price of electricity in your area and your use of the refrigerator.

New national efficiency standards for refrigerators and freezers will take effect in 1993. When this book went to press, only a handful of models met, or were close to meeting, those new standards. The models listed here may be joined or replaced by many other efficient models during 1993 and beyond.

Chill Out

In 1992, a consortium of utilities announced a $30 million bounty to the company that could produce a highly energy-efficient refrigerator. In the mean time, it should be noted that there are things you can do with your existing model to cut its energy use. One of those things comes in the form of a new product called ChillShield, a rather low-tech device that reduces cold-air loss when the door is opened. ChillShield consists of two layers of overlapping 3"-wide vinyl flaps along with a special mounting system that allows a customized fit to any size or type of refrigerator. Mounted inside the refrigerator, the flaps easily part when you need to reach in to get something, then fall back into place, forming an airtight barrier that keeps cold air in. According to its manufacturer, ChillShield can cut energy use by up to 20 percent. Assuming energy rates of 10 cents per kilowatt-hour, says the manufacturer, ChillShield could cut the average household energy bill by $30 a year, which will cover the product's $25 purchase price in about 10 months. (There's also a $35 model that includes a second shield for the freezer compartment.) For more information contact The Conserve Group, P.O. Box 1560, Bethlehem, PA 18016; 215-691-8024.

Top Freezer, Automatic Defrost, 14.0–16.4 Cubic Feet

Brand	Model	Volume	Energy Use	Energy Cost
Frigidaire	MRT*15CHA**	15.0	563	46
Gibson	MRT15CHA**	15.0	563	46
Tappan	MRT15CHA**	15.0	563	46
Kelvinator	MRT15CHA**	15.0	563	46
White-Westinghouse	MRT15CHA**	15.0	563	46

Top Freezer, Automatic Defrost, 16.5–18.4 Cubic Feet

Brand	Model	Volume	Energy Use	Energy Cost
Frigidaire	MRT17CHA**	16.8	639	53
Gibson	MRT17CHA**	16.8	639	53
Tappan	MRT17CHA**	16.8	639	53
Kelvinator	MRT17CHA**	16.8	639	53
White-Westinghouse	MRT17CHA**	16.8	639	53

Top Freezer, Automatic Defrost, 14.0–16.4 Cubic Feet

Brand	Model	Volume	Energy Use	Energy Cost
Frigidaire	FPES19TS**	18.6	722	60
KitchenAid	KTRS20*X**2*	19.9	725	60
Whirlpool	ET20G**Z*0*	19.9	765	63
Kenmore	9*005*6	19.9	765	63
Kenmore	9*103*3	19.9	765	63

Top Freezer, Automatic Defrost, 20.5–22.0Cubic Feet

Brand	Model	Volume	Energy Use	Energy Cost
General Electric	TBH*21ZR	20.6	734	61
Hotpoint	CTH21ER	20.6	734	61
Kenmore	9*618	20.6	734	61
KitchenAid	KTRS22*X**2*	21.6	801	66
Whirlpool	ET22*M*Z*0*	21.7	802	66
Kenmore	96082*2 21.7	802	66	63

Side-by-Side, Automatic Defrost, 19.0–21.9 Cubic Feet

Brand	Model	Volume	Energy Use	Energy Cost
Frigidaire	FRS20*RA**	19.6	783	65
Frigidaire	FPI20VS**	19.6	852	70
Kenmore	84902*6	19.8	856	71
Roper	RS20CK*Z*0*	20.0	856	71
Whirlpool	ED20*K*Y*0*	19.9	856	71

Tips for Refrigerator Efficiency

• **Keep the condenser coils clean.** (The coils are on the back or at the bottom of the refrigerator.) At least once a year, carefully wipe, vacuum, or brush the coils to remove dust and dirt. Many refrigerators provide access to coils through a removable panel.

• **Keep the door gasket clean and tight.** This will ensure that no cold air is escaping into the room from the refrigerator. To check the tightness, close the refrigerator door with a single sheet of paper placed between the door and the cabinet. With the door closed, try pulling on the paper. It should come loose with a bit of effort. If it comes loose too easily, you may need to replace the gasket or adjust the hinges. Do this test in several places along the door.

• **Check the temperature.** You may wish to keep a thermometer in the refrigerator to check the temperature periodically. The ideal temperature for the refrigerator is between 38 degrees and 40 degrees Fahrenheit; the freezer ideally should be 0 degrees Fahrenheit.

FREEZERS

The energy efficiency of freezers has improved over the past decade, although at a slower rate than refrigerators.

Top Freezer, Automatic Defrost, 14.0–16.4 Cubic Feet

Brand	Model	Volume	Energy Use	Energy Cost
General Electric	CB15**	14.8	454	37
Estate	TCF15*	14.8	454	37
Crosley	C15HS	14.8	454	37
Admiral	C15HS	14.8	454	37
Wood's	CWC15-ZX	14.8	431	36

Dishwashers

Most of the cost of running dishwashers (and clothes washers) is for heating water. So, an efficient water heater is essential to reducing energy use. The annual energy costs shown are based on a typical electric water heater and an electricity price of 7.63 cents per kilowatt-hour, the price used on the most recent Energy Guide labels for dishwashers.

Standard Size (capacity for 8 or more place settings)

Brand	Model	kWh/Yr.	Energy Cost
Miele	G5**SC	655	52
KitchenAid	KUD-23 Builder	649	54
KitchenAid	KUD-23 DY	646	53
Miele	G572*	630	52
Miele	G590*	630	52
A.E.G.	Favorit 665i	520	43

Clothes Washers

As with dishwashers, the biggest cost is hot water. The annual energy costs shown are based on a typical electric water heater and an electricity price of 7.63 cents per kilowatt-hour, the price used on the most recent Energy Guide labels for clothes washers.

Clothes Washers

Brand	Model	kWh/Yr.	Energy Cost
White-Westinghouse	LT200P	291	24
White-Westinghouse	LT350R	291	24
Kenmore	4988*	291	24
Gibson	WS27M8WA	291	24
A.E.G.	Bella Super	388	32
A.E.G.	539	388	32

Water Heaters

In recent years, new types of water heaters have come onto the market. In addition to the traditional storage water heater—in which a ready reserve of water is always kept hot—a new type of "demand" water heater heats water only when it is needed. While hot water never runs out, the rate of flow is limited. This trade-off results in an energy savings of about 25 percent, or about $75 per year for gas appliances, a bit more for electrical appliances.

There are other energy-saving water heater technologies worth considering. Some newer systems are designed to take advantage of a home's primary heating system to heat hot water. Solar water heaters use the sun's heat to produce hot water, often in combination with a backup conventional heating unit during cold periods with little sunshine. Heat-pump water heaters take heat from the air and move it to the water in a storage tank. Each of these has advantages (they are energy efficient and less expensive than traditional water heaters) and disadvantages (they don't always get water as hot as traditional water heaters) and are worth considering if you plan to install a new heater.

The energy efficiency of a water heater is indicated by its Energy Factor, or "EF," an overall number based on the use of 64 gallons of hot water per day. The average new gas water heater sold in 1988 had an EF of about 0.85. Under the new federal standards, the minimum EF ratings for today's water heaters are based on storage tank size. In general, the smaller the tank size, the higher the EF. (Higher numbers indicate better efficiency.)

Minimum Energy Factors for Water Heaters

Tank Size	Gas	Oil	Electric
30 gallons	0.56	0.53	0.91
40 gallons	0.54	0.53	0.90
50 gallons	0.53	0.50	0.88
60 gallons	0.51	0.48	0.87

Choosing the right tank size is very important. If it is too large, you will waste a lot of energy heating water you'll never use. If it's too small, you're bound to run out of hot water in the middle of your shower. The ability of hot water heaters to meet peak demands for hot water is indicated by the "first hour rating" (listed in the "rating" column below). This rating accounts for the effects of tank size and the speed by which cold water is heated. In some cases a water heater with a small tank but a powerful burner or heating element can have a higher first-hour rating than a unit with a large tank and a less-powerful burner or heating element. (For more information on first hour ratings, contact the **Gas Appliance Manufacturers Association**, 1901 N. Moore St., Arlington, VA 22209; 703-525-9565.) The models listed on the next pages are grouped by their tank size.

Electric Water Heaters, 30 gallon tanks

Brand	Model	1st hr. rating (gal.)	Storage Tank (gal.)	Energy Factor
Marathon	MP30255	46	30	.96
Sears	449.320310	46	30	.96
Marathon	MP30245	42	30	.96
Reliance	LT 30 2LRT6-W	41	30	.96
State	TCL 30 2LRT6-W	42	30	.96

Oil-Fired Water Heaters, Approx. 30 gallon tanks

Brand	Model	1st hr. rating (gal.)	Storage Tank (gal.)	Energy Factor
Bock	32PP	131	32	.65
Bock	30ES	109	28	.64
GSW	JW3*7	120	30	.62
Bradford-White	F-I-305E50EZ	120	30	.61
Ford	FG3016EZ	120	30	.61

Gas-Fired Water Heaters, Approx. 30 gallon tanks

Brand	Model	1st hr. rating (gal.)	Storage Tank (gal.)	Energy Factor
American Appliance	DVPB35	147	34	.86
American Appliance	N2F363T	61	30	.64
Bradford-White	M-III-303T5CN-7	64	29	.64
Craftmaster	30	61	30	.64
Rheem	21XR30-5*	52	30	.64

Room Air Conditioners

An air conditioner takes heat from a room, using a special heat-absorbing fluid called a refrigerant (usually containing Freon or another type of chlorofluorocarbon), and moves the heat outside. As heat is removed from the room, the room becomes cooler.

In choosing a room air conditioner, you should consider two primary factors: the cooling capacity and the operating efficiency. The cooling capacity is expressed in BTU/hour; your appliance dealer should provide a Cooling Load Estimate Form from the Association of Home Appliance Manufacturers showing the correct cooling capacity for the room you plan to cool.

The operating efficiency of room air conditioners is rated by their Energy Efficiency Ratio, or EER, the ratio of the cooling output divided by the power consumption. The higher the EER, the more efficient the air conditioner. An average new room air conditioner has an EER of about 8. A model with an EER greater than 8.5 is considered "efficient"; a model with an EER over 9.5 is "very efficient." The new national standards set the minimum average EER at about 8.6.

Room Air Conditioners—Less Than 6,000 BTU/Hr.

Brand	Model	Volts	Capacity	EER
Airtemp	C3R06F2B	115	5800	10.0
Climatrol	M3R06F2B	115	5800	10.0
Fedders	A3R06F2B	115	5800	10.0
Marta America	W3R06F2B	115	5800	10.0
Micro Sonic	3S062A	115	5800	10.0

Room Air Conditioners—6,000–7,999 BTU/Hr.

Brand	Model	Volts	Capacity	EER
Friedrich	SS07H10A	115	7200	11.0
Friedrich	SQ06H10	115	6700	10.3
Cold Point	CP08SA	115	7800	10.2

Room Air Conditioners—8,000–9,999 BTU/Hr.

Brand	Model	Volts	Capacity	EER
Cold Point	CP08SA	115	7800	10.2
Friedrich	SQ06H10	115	6700	10.3
Cold Point	CP08SA	115	7800	10.2
Cold Point	CP08SA	115	7800	10.2

Room Air Conditioners—10,000–13,999 BTU/Hr.

Brand	Model	Volts	Capacity	EER
Cold Point	CP10SAE	115	10,500	12.1
Friedrich	SM10H10A	115	10,300	12.0
Panasonic	CW-102VS12L6M	115	10,000	11.6
Quasar	HQ2102CW	115	10,000	11.6
Amana	14C2MA	115	13,700	10.4
Cold Point	CP12SAE	115	11,800	10.4

Room Air Conditioners—14,000–16,999 BTU/Hr.

Brand	Model	Volts	Capacity	EER
Freidrich	SM14H10A	115	14,000	10.5
Carrier	51GMB11411	115	14,000	10.2
Kenmore	106.8761492	115	14,000	10.2
Hampton Bay	BHAC1400XS0	115	14,000	10.2

Room Air Conditioners—17,000–19,999 BTU/Hr.

Brand	Model	Volts	Capacity	EER
Friedrich	*L19H3*B	230/208	19,000	10.0
Airtemp	C3L19E7A	230/208	18,500	10.0
Climatrol	M3L19E7A	230/208	18,500	10.0
Fedders	A3L19E7A	230/208	18,500	10.0
J. C. Penney	8572067	230/208	18,500	10.0

Central Air Conditioners

Central air-conditioning (CAC) units operate similarly to room air conditioners, but they are rated differently. The standard is the Seasonal Energy Efficiency Rating, or SEER, which is the cooling output divided by the power input for an average U.S. climate. The average CAC sold in 1988 had a SEER of about 9; many older units have a SEER of 6 or 7. The 1992 national efficiency standards call for a minimum SEER of 10. In general, a SEER of 10 is good; a SEER of 12 is considered excellent.

If you have an older, inefficient unit, do not replace its compressor unit with a newer, more efficient one unless it is compatible with the CAC's blower coil. A highly efficient compressor may not achieve its rated efficiency if paired with an older blower coil. Ask your appliance dealer or service technician for guidance.

The air conditioners below are grouped by their cooling capacity. Each "ton" represents 12,000 BTU/hour of cooling capacity.

Central Air Conditioners
Cooling Capacity: Approx. 2 Tons

Brand	Condensing Unit	Blower Coil	BTU/Hr	SEER
Carrier	38TV02430	28RVS017+58SSC0**+38T V900001+HC44AE115	24,600	16.00
American Standard	ATX024C	TWH039E15-C	26,200	14.40
Trane	TTX024A	TWH039E15-C	26,400	14.40
Carrier	38TR02430	FK4A-002	24,400	14.00

Central Air Conditioners
Cooling Capacity: Approx. 2.5 Tons

Brand	Condensing Unit	Blower Coil	BTU/Hr	SEER
Trane	TTS730A	THD080A9V3+TXF736S5	32,000	15.85
Trane	TTS730A	T*C080B9V3+TXC736S5	30,800	15.70
American Standard	ATX030C	TWH039E15-C	32,800	14.20
Trane	TTX030A	TWV039E15-C	32,800	14.20

Maintaining Heating and Cooling Systems

Air conditioners and heat pumps need regular maintenance to function at peak efficiency:

• **Room air conditioners:** Inspect the filter monthly and change or clean it. The condenser should be cleaned professionally every few years. During the winter, make sure the air conditioner or its opening is sealed tightly to keep out winter drafts.

• **Central air conditioners and heat pumps:** They should be inspected, cleaned, and tuned every three years by a professional service person. Besides extending the life of the system, regular servicing can lead to 10 to 20 percent energy savings. The power to a CAC should be turned off during the winter; otherwise, the heating elements will consume power all winter long. To turn the power off, flip the circuit breaker by the outside unit of the air conditioner. The power should be turned on at least one day before the air conditioner is actually used, to prevent compressor damage.

Central Air Conditioners
Cooling Capacity: Approx. 3 Tons

Brand	Condensing Unit	Blower Coil	BTU/Hr	SEER
Trane	TTS730A	TW*739E15-C	33,200	16.90
Carrier	39TV03630	28RVS017+58SSC075+38TV900001+HC44AE115	36,000	16.00
Carrier	39TV03630	40QL01730	36,000	16.00
Bryant	598A036	FK4ANF006	40,000	16.00
Day & Night	598A036	FK4ANF006	40,000	16.00

Central Air Conditioners
Cooling Capacity: Approx. 3.5 Tons

Brand	Condensing Unit	Blower Coil	BTU/Hr	SEER
Carrier	38TD03630	FK4ANF006	40,000	16.60
Trane	TTS736A	TW*739E15-C	39,000	16.20
Lennox	HS14-411V-7P	CB*21-51-1P+LB-53081CA	43,000	15.50
Lennox	HS14-411V-7P	CB*21-41-1P+LB-53081CA	39,500	15.00

Central Air Conditioners
Cooling Capacity: Approx. 4 Tons

Brand	Condensing Unit	Blower Coil	BTU/Hr	SEER
Trane	TTS748A	TW*064E15-C	49,500	15.15
Carrier	38TD04830	FK4ANF006	51,000	14.90
Bryant	598A048	FK4ANF006	51,500	14.50
Day & Night	598A048	FK4ANF006	51,500	14.50
Payne	598A048	FK4ANF006	51,500	14.50

Central Heat Pumps

Heat pumps have two modes: heating and cooling. In the cooling mode, the heat pump works just like an air conditioner. In the heating mode, it works on the principle that even cool air contains some heat, and this heat may be extracted and concentrated, providing warm air to heat your home. In the heating mode, a heat pump uses a refrigerant to absorb heat from the cold outside air. A compressor concentrates the refrigerant, which causes the air's temperature to rise. The warmed air is then transferred indoors. Because they do not perform well over extended periods of subfreezing temperatures, heat pumps are generally most cost effective in regions without severe winters.

Like central air conditioners, the cooling ability of heat pumps is rated by a SEER; the heating ability is rated by the Heating Season Performance Factor, or HSPF, a ratio of the estimated seasonal heating output divided by the seasonal power consumption for an average U.S. climate. A typical new heat pump has an HSPF of about 6.5 and a SEER of about 9. The national energy-efficiency standards call for a minimum HSPF of 6.8 and a minimum SEER of 10.

Central Heat Pumps
Capacity: Approx. 1.5 Tons

Brand	Outdoor Unit	Indoor Unit	BTU/hr. Cooling	SEER	BTU/hr. Heating	HSPF
Coleman	3718A911	V17-30KV	18,600	12.00	20,000	8.00
Lennox	HP22-211-1P	CB19-21-1FF+LB3 4792BE	19,800	12.85	19,000	7.90
Lennox	HP19-211-1P	CB19-21-*P+LB34 792BE	19,600	12.00	19,200	7.80
Coleman	3718-611	6930-CUV0	19,400	11.00	19,900	7.80
Bryant	697BN018-A	FK4ANF002	18,500	13.00	18,500	7.80

Central Heat Pumps
Capacity: Approx. 2 Tons

Brand	Outdoor Unit	Indoor Unit	BTU/hr. Cooling	SEER	BTU/hr. Heating	HSPF
Carrier	38YV02430	40YVM00530	24,800	15.50	25,400	10.00
Carrier	38QE92430	38QE023430+40 QE02430	24,000	13.35	25,800	8.75
Century	HS1324-1A	AM08+SHA1324-U	24,800	13.10	25,200	8.75
HeatController	HS1324-1A	AM08+SHA1324-U	24,800	13.10	25,200	8.75
Rheem	RPGB-025JA	RHQA-08+RCQB-B0 24	24,800	13.10	25,200	8.75

Central Heat Pumps
Capacity: Approx. 2.5 Tons

Brand	Outdoor Unit	Indoor Unit	BTU/hr. Cooling	SEER	BTU/hr. Heating	HSPF
Coleman	3730A911	3736-833	31,000	13.80	31,400	9.00
Trane	TWS730A	TWV739E15-C	31,800	16.40	29,600	8.70
Lennox	HP22-311-*P	CB19-31-*P+LB-34 792BG	31,000	13.40	31,200	8.65
Trane	TWX030C	TWV039E15-C	31,200	13.10	27,600	8.5

Central Heat Pumps
Capacity: Approx. 3 Tons

Brand	Outdoor Unit	Indoor Unit	BTU/hr. Cooling	SEER	BTU/hr. Heating	HSPF
Carrier	38YV03630	40YVM00530	35,600	15.50	35,600	10.00
Carrier	38QE93630	38QE03630+	36,800	14.05	35,400	9.05
Coleman	3736A911	3736-833	37,000	13.50	37,000	9.00
Trane	TWS736A	TWV739E15-C	37,600	15.20	34,600	8.75

Central Heat Pumps
Capacity: Approx. 3.5 Tons

Brand	Outdoor Unit	Indoor Unit	BTU/hr. Cooling	SEER	BTU/hr. Heating	HSPF
Bryant	698AN036-A	FK4ANF006	40,000	15.80	38,000	8.80
Day & Night	698AN036-A	FK4ANF006	40,000	15.80	38,000	8.80
Payne	698AN036-A	FKANF006	40,000	15.80	38,000	8.80
Lennox	HP21-511-*P	CB21-51-1P+LB-3 4792BF	44,500	16.00	41,000	8.75
Trane	TWS748A	TDC100B9V5+TXC 0,754S	42,500	13.75	42,500	8.70

New, Improved Heat Pumps

Mention heat pumps and most people think of chilly winters and sweaty summers. And with good reason: The heat pumps of just a few years ago didn't do a very good job.

That's changed. Experts say that the newer generation provide all the benefits heat pumps have to offer—affordable, energy-efficient heating and air conditioning—without the discomfort. So, today's heat pumps are worth looking at if you are considering upgrading your home's heating and cooling system.

Heat pumps work like this: In winter, they draw in outside air and blow it over a coil filled with freon, a refrigerant. That refrigerant absorbs the heat in the air. When the refrigerant is heated, it becomes a low-temperature hot gas. That gas is then compressed, which makes it much hotter, about 180°, heating a series of coils. A fan blows air over the coils, and that heated air is circulated through ducts. In the summer, the process is reversed, with the refrigerant extracting hot air from inside, then cooling it and blowing the cooled air through ducts in your home.

The only energy used by heat pumps is the electricity needed to drive a compressor and a fan to blow the heated or cooled air through your home. The exception is in colder climes: When it gets too cold, heat pumps can't extract enough heat from the air so a traditional electric resistance heater must kick on, raising the temperature... and your electric bill.

At least that's the case with an *air-source* heat pump, which works best in moderate climates and in the South, where there's a need for a lot of cooling in the summer.

A newer development is the *ground-source* heat pump. This type exchanges heat or cool through a closed loop of pipes buried in the earth, which remains at a much more constant temperature—typically, between 42° and 77°—than the air. That makes this type of heat pump much more efficient than an air-source unit, but the installation cost is higher; among other things, it requires some digging in your yard. It's much easier to install these models during new construction than in existing homes.

In general, heat pumps cost more to buy and install than conventional furnaces and air conditioners. The savings come in the operating costs, making them cheaper to operate over the long run. A typical ground-source heat pump system will average between 10 percent and 30 percent over a fossil fuel heating system with central air conditioning.

There's one more environmental consideration about heat pumps: though they may be more energy efficient, resulting in fewer emissions into the atmosphere, there's the matter of the refrigerants they use. In most air-source models, it is freon, the stuff that's eating away the ozone layer. In ground-source units, R-22, a hydrochlorofluorocarbon, or HCFC, is used. HCFCs are only about 5 percent as destructive to ozone as CFCs.

If you're planning to upgrade your heating system, heat pumps can make a lot of sense.

Gas Furnaces

New furnaces are rated by their Annual Fuel Utilization Efficiency, or AFUE, a measure of overall seasonal performance. The average new gas furnace sold in 1988 had an AFUE of 75 percent. The 1992 national efficiency standards call for a minimum AFUE of 78 percent.

The efficiency of a gas furnace has a lot to do with the sophistication of the equipment. There are five major systems:

Type of furnace	Typical AFUE
Gas pilot light	62 percent
Electronic ignition	68 percent
Automatic vent damper	76 percent
Power combustion	82 percent
Condensing furnace	90 percent+

Condensing furnaces are often priced much higher than less-efficient types of furnaces, making them most economical in areas with long, cold winters. Those in regions with moderate or mild winters may suffice with an AFUE in the 78 percent to 85 percent range. Of course, one important energy-efficiency feature is to make sure that the furnace capacity is appropriate for your home.

Gas Furnaces—26,000–42,000 BTU/Hr.

Brand	Model	BTU/Hr.	AFUE (%)
BDP	398AA*030040	40,000	96.6
Carrier	58SX*040-FG	40,000	96.6
Lennox	G21Q3-40-	38,000	96.2
Armstrong	EG6E40DC13	38,000	94.5
DMO Industries	HCS2-40	38,000	93.7

Gas Furnaces—43,000–59,000 BTU/Hr.

Brand	Model	BTU/Hr.	AFUE (%)
Metzger	HBCG-60	56,000	96.0
Thermo-Products	GLC-50	48,000	96.0
Amana	GUX045B30A	43,000	95.5
Lennox	GSR21Q4-50	47,000	95.3
DMO Industries	HCS2-40	38,000	93.7

Gas Furnaces—60,000–76,000 BTU/Hr.

Brand	Model	BTU/Hr.	AFUE (%)
Metzger	HBCG-80	75,000	95.5
Amana	GUX070B30A	67,000	95.3
Amana	GUX070B40A	67,000	95.1
Thermo-Products	GLC-75	71,000	95.0
Amana	EGHW100DC3	76,000	94.9

Gas Furnaces—77,000–93,000 BTU/Hr.

Brand	Model	BTU/Hr.	AFUE (%)
Amana	GUX090B50A	87,000	95.5
Amana	GUX090B35A	86,000	95.3
Metzger	HBCG-100	93,000	95.0
Lennox	G21Q3-100	93,000	94.0
Amana	EGHW100DC3	93,000	93.9

Gas Furnaces—94,000–110,000 BTU/Hr.

Brand	Model	BTU/Hr.	AFUE (%)
Amana	GUX115B50A	110,000	95.0
Lennox	G21Q4/5-100-	95,000	94.5
Heat Controller	GZUB105-E4N	100,000	93.8
Nordyne	G1RC-105A19	100,000	93.8
Rheem	RGEC-10ERAJ*	100,000	93.8

Oil Furnaces

The 1992 national efficiency standards for oil furnaces are the same as for gas furnaces: a minimum of 78 percent. High-efficiency oil furnaces achieve their high ratings through automatic flue dampers and "flame retention" burners.

Oil Furnaces—43,000–59,000 BTU/Hr.

Brand	Model	BTU/Hr.	AFUE (%)
Enerroyal	ERHB56 SF	58,000	86.3
Yukon	85-100-1500	58,000	86.2
Enerroyal	ERLB56 SF	58,000	86.0
Thermo-Products	OL2-56-V	57,000	85.6
Oneida Royal	LOR56A-3,3B	59,000	84.9

Oil Furnaces—60,000–76,000 BTU/Hr.

Brand	Model	BTU/Hr.	AFUE (%)
Coleman	M99-H70/90	64,000	88.9
Yukon	U-70-0-03	64,000	88.9
Coleman	85M	63,000	87.8
Yukon	H-70-0-02	63,000	87.8
Enerroyal	ER75	75,000	86.7

Oil Furnaces—77,000–93,000 BTU/Hr.

Brand	Model	BTU/Hr.	AFUE (%)
Coleman	M99-70/90	83,000	89.1
Yukon	U-90-0-03	83,000	89.1
Bard	FC085D36A	77,000	86.6
Bard	FC085D36A	88,000	86.1

Oil Furnaces—94,000–110,000 BTU/Hr.

Brand	Model	BTU/Hr.	AFUE (%)
Thermo-Products	OL11-105M	101,000	86.0
Bard	FLH110D48A	97,000	85.7
Enerroyal	ERLB95 SF	98,000	85.0
Metzger	*BO-120	100,000	85.0
Thermo-Products	OH11-105-V	101,000	84.6

PRODUCTS TO SAVE ENERGY

In addition to major appliances and lighting, here are several smaller, inexpensive products that can save you energy and money.

• **Air-conditioner covers:** If you cannot remove your room air conditioner from the window during the winter, consider covering it, both inside and out. Besides protecting your air-conditioning unit, these covers also help keep cold air from entering your home through the space around the air-conditioner, thus cutting down on heating costs. One brand is **Frost-King Air-Conditioner Cover**.

• **Caulking:** Filling in the small spaces and gaps around window openings and where pipes and wires enter the home reduces drafts that impede the efficiency of your heating and air-conditioning system. Most caulking products cost under $10; rope caulk, one of the easiest types to apply, sells for about $4 for 40 or 50 feet. There are many brands available, sold at most hardware stores, including **DAP Rely-ON, Dow-Corning Silicone Caulk, Easy Caulker Acrylic Foam,** and **Mortite Rope Caulk**.

• **Draft Blockers:** These foam plates fit behind light switches and electrical outlets to reduce drafts that come in through those spaces. You can get a packet of 10 for about $3; they are easy to install, with only a screwdriver.

• **Draft Guards:** These are sand- or pebble-filled tubes that you place around windowsills and sashes to reduce the loss of heated and cooled air. (If you prefer to sleep with the window open in winter, placing a draft guard under your bedroom door will prevent the cold air from seeping into the rest of the house.) Draft guards cost about $7 and are sold under a variety of names, including **Frost-King Stop Draft Weatherstrip**.

• **Heat Reflectors:** These are thin sheets (usually foam coated on one side with metal foil) that fit behind radiators, to reflect heat away from the wall and into the room, thereby maximizing each radiator's efficiency. The **Viking Heat Reflector** is one such product.

• **Programmable Thermostat:** These new types of thermostats allow you to change the temperature (of both heating and air conditioning) at different times of day. You might, for example, have the heater go to 68 degrees at 6 A.M. (to have everything warm by the time

Keeping Current

There's no such thing as too much good information on saving energy. Our nation's inefficiency in this area remains a top environmental threat and when someone as authoritative as the Rocky Mountain Institute (RMI) publishes a handbook with energy-saving advice, it's worth getting. What makes *Practical Home Energy Savings* especially worthy is its very readable style on this not so breezy subject. Here's a sampling:

• Make tightening the skin of your home or apartment the first priority. A typical house in the U.S. has about five square feet of air leaks, which account for 30-40 percent of the heating and cooling bill.

• A modern dishwasher will save energy and water, but only if you plunk the dishes straight in without pre-washing them first.

• Tanks to modern technology, most televisions draw from 1.5 to 8 watts even when "off." That's so they'll be warmed up to turn on instantly at the flick of a switch.

• The best way to maintain indoor air quality isn't through air conditioning. It's by carefully choosing the materials, cleaners, and carpets that go into your home to minimize the source of pollutants.

• A single three-foot-by-five-foot window in direct summer sun can let in more light than 100 standard 60-watt bulbs.

In the end, say the authors, if Americans were as energy-efficient as western Europeans or Japanese, we could save $200 billion a year—roughly enough to wipe out the federal deficit or put an extra $800 in the pocket of every man, woman, and child.

Practical Home Energy Savings is available for $8 from RMI, 1739 Snowmass Creek Rd., Snowmass, CO 81654; 303-927-3851.

you get up), drop to 60 degrees at 9:30 A.M. (after everyone has left for the day), go back up to 68 degrees at 4 P.M. so the house will be warm when you return for the evening), then drop back to 60 degrees at 11 P.M. (while you are in bed). Some programmable thermostats also have a second set of settings for weekends, when people are often at home more during the day. The thermostats, which range from $90 to $175, are not cheap, but they can save up to 12 percent on your heating bill each year. The device surely will pay for itself within two or three years. There are many brands, available at most hardware stores, including **White-Rogers Comfort-Set, Sears Thrift-T, Honeywell Fuel Saver,** and **Weather Wizard.**

• **Reflective Window Film:** These are thin plastic sheets you place directly on the inside of window panes and glass doors. The film reflects inside heat back into your home, reducing the amount that is conducted outside through windows. The film, which costs about $10 a window, is easy to put on; it adheres to the window directly, or with the help of water from a spray bottle. One good brand is **3M Scotchtint Reusable Sun Control Film.**

• **Storm Window Kits:** It can be expensive to have storm windows installed throughout your house (although the investment will probably pay for itself in fuel savings and increased resale value of your home), but there is a less-expensive alternative. Storm window kits consist of plastic film or sheets to cover the window. Attaching the plastic to the window is done with tape or tacks. One product, **3M Window Insulation Kit**, includes "shrink" film that you apply by using the hot air from a hair dryer to tighten the film and shrink away wrinkles. Other brands include **Frost-King Storm Window Kit, Sears Window Insulating Kit,** and **TYZ All-Weather Window**. Prices range from about $3 to $10 per window.

• **Weatherstripping:** This includes plastic, foam, felt, or rubber strips that fit around window and door frames to create a tight seal and reduce heated and cooled air from escaping outside. Most are easy to apply, usually by way of a self-adhesive backing. Prices vary, but average about $5 per window or door. Brands include **Frost-King Weatherstrip Tape, 3M V Seal Weatherstrip, Jalousie Window Weatherstrip, Stix-on Self-Adhesive Weatherstripping,** and **Mortite Weatherstripping and Caulking Cord**.

Where to Buy Energy-Saving Products

The charts on the next two pages illustrate the widespread availability of many energy-saving and other environmentally conscious products. As you can see, most of the types of products discussed in this chapter can be found in the major hardware and home-furnishing chains—as well as in many, many smaller chains and independent stores. In addition, several of the leading mail-order companies feature a wide array of products. You are encouraged to obtain the latest catalogs from these companies. In addition to showcasing products, some catalogs provide a wealth of educational material on home-energy and related topics.

Where to Find Environmentally Conscious Products

Major Retailers

	Ames	Ace Hardware	Builders Square	Hechingers	Home Depot	K-mart	Lowe's	Meijer	Montgomery Ward	Payless Cashways	Sears, Roebuck	Target	True Value	Wal-Mart
Can crushers		•	•	•	•	•	•	•	•	•	•	•	•	•
Cedar closet liner/blocks		•	•	•	•	•	•	•	•	•	•	•	•	•
Chipper/shredders			•	•	•		•	•		•	•	•	•	•
Compact fluorescent bulbs	•	•	•	•	•	•	•	•		•	•	•	•	•
Compost bins		•	•	•	•	•	•	•		•	•	•	•	•
Enzyme drain cleaners		•	•	•	•	•	•	•		•	•		•	•
Faucet aerators		•	•	•	•	•	•	•		•	•	•	•	•
Insecticidal soaps				•	•	•	•			•		•	•	•
Low-flow showerheads		•	•	•	•	•	•	•		•	•	•	•	•
Low-water-use toilets			•	•	•		•			•	•	•	•	
Mulching lawn mowers		•	•	•	•	•	•	•	•	•	•	•	•	•
Non-halon fire extinguishers	•	•	•	•	•		•	•		•	•	•	•	•
Organic fertilizers				•	•	•	•	•		•	•	•	•	•
Programmable thermostats		•	•	•	•		•	•		•	•		•	
Recycled platic trash bags		•	•			•		•		•	•	•	•	•
Recycling bins	•	•	•	•	•	•	•	•	•	•	•	•	•	•
Soaker hoses		•	•	•	•	•	•	•		•	•	•	•	•
Toilet leak repair kits		•	•	•	•	•	•	•		•	•	•	•	•
Water timers		•	•	•	•	•	•	•		•	•	•	•	•
Weatherstripping	•	•	•	•	•	•	•	•		•	•	•	•	•

Where to Find Environmentally Conscious Products

Mail-Order Companies

	Energy Store	Jade Mountain	Real Goods	Seventh Generation
Can crushers			•	
Cedar closet liner/blocks			•	•
Chipper/shredders				
Compact fluorescent bulbs	•	•	•	•
Compost bins	•		•	•
Enzyme drain cleaners		•		•
Faucet aerators	•	•	•	•
Insecticidal soaps				
Low-flow showerheads	•	•	•	•
Low-water-use toilets		•	•	
Mulching lawn mowers			•	
Non-halon fire extinguishers				
Organic fertilizers				•
Programmable thermostats	•		•	
Recycled platic trash bags			•	•
Recycling bins			•	
Soaker hoses				
Toilet leak repair kits			•	
Water timers		•	•	
Weatherstripping			•	

Here are telephone numbers for the catalogs mentioned in this chart: **The Energy Store:** 800-288-1938; **Jade Mountain:** 800-442-1972; **Real Goods Trading Co.:** 800-762-7325; and **Seventh Generation:** 800-456-1177.

GREEN LIGHTS

How many light bulbs does it take to screw up the environment? The question is more than a mere mutation of an old riddle. Fully one-fifth of all electricity used in the United States is for lighting; at home, lighting typically consumes about a tenth of every energy dollar. Generating the electricity to burn all those lights contributes to a host of environmental problems.

Cutting this energy load offers benefits for everyone. For the environment, it means fewer emissions from coal-fired generating plants that contribute to global warming and acid rain, and radioactive wastes from nuclear plants. For utility companies, it could mean building fewer power plants. For consumers, it could save money—and make the planet a little less polluted. Sounds pretty good, right?

The solutions, unfortunately, aren't always as simple as removing one light bulb and replacing it with another. There are trade-offs galore. The good news is that there are a fast-growing number of energy-saving lighting options available, and many fine products worth looking into.

But beware: It's important to keep in mind that there is no single "best" light bulb. The right light depends upon where and how it will be used, among other considerations.

THE RIGHT LIGHT

As with other household "appliances," you probably don't realize that the cost of a light bulb is only a fraction of its total cost. The "operating cost"—the amount you pay for electricity—is five to ten times the cost of the bulb itself. So, those two bulbs you are buying for $2.50 actually cost you $12.50 to $25 in electricity.

To understand how to save money, it may first help to understand the three basic types of household lighting.

• **Incandescent:** This type—the standard light bulb invented in the 1800s—is the most common lighting source in American homes. It is also the most wasteful. Ninety percent of the energy consumed by an incandescent lamp is given off as heat rather than visible light. This is because incandescents do not directly convert electricity to light.

Rather, they use electricity to heat a coiled filament in a vacuum or inert gas-filled bulb, making the filament glow.

• **Fluorescent:** Fluorescent lamps—usually sold as long, thin tubes with two prongs at each end—convert electricity to visible light by using an electric charge to "excite" gaseous atoms. These atoms emit ultraviolet radiation, which is absorbed by the phosphor coating on the tube walls. The phosphor coating "fluoresces," producing a visible light. Fluorescent lamps convert electricity to visible light up to five times more efficiently than incandescent lamps, and last up to twenty times longer. Most fluorescent lamps require a special fixture, although a new generation of compact fluorescent lamps can fit in existing "screw-in" fixtures.

• **High-intensity discharge (HID):** The most common type found in homes is halogen, found increasingly in reading lights and some overhead fixtures. These lamps vary from the standard incandescent in that a halogen gas is added to the bulb. Halogen gas reduces the filament evaporation rate, thereby reducing the rate at which the glass bulb darkens. This increases lamp life up to four times that of a standard bulb. Moreover, tungsten halogen lamps emit more light for the same amount of electricity. These bulbs require special fixtures, however, and can become very, very hot while lit. The bulbs are also very expensive compared to standard incandescent or fluorescent bulbs.

A New Breed of Bulb

Chances are you've heard a thing or two about compact fluorescent (CF) lights, a new breed of bulb with tremendous energy-saving potential. CFs are an improved variation on those long fluorescent tubes that flicker and buzz and give some people headaches at work or in supermarkets.

But manufacturers of CFs, which are designed to screw into the same sockets as traditional incandescent bulbs, have eliminated some of those problems. For starters, the bulbs no longer flicker or buzz. Moreover, the quality of light they emit is far more appealing than the "cool-white" tubes at work.

Still, CFs aren't the lighting cure-all some people believe them to be. For one thing, they don't fit in all types of light fixtures. They can't be used with dimmers. They aren't suitable for lights that are turned on for only a few minutes at a time, such as those in bathrooms, closets, and garages. And they are expensive to buy, though they save money over the long run.

That's not all. There are other issues surrounding CFs that we'll get into in a moment. But let's first deal with the matter of cost.

CFs typically cost about $18 to $20 to buy, considerably more than the 75 cents to a dollar price for an incandescent bulb. How can anything so relatively expensive save money?

In two ways. By using only about one-fourth the energy, and lasting up to ten times longer than incandescents, CFs reduce electricity costs and the price of replacement bulbs.

The numbers work like this. Let's be conservative and assume the CF costs $20 and the incandescent 75 cents. Assume the CF uses only 18 watts to give the same amount of light as a 60-watt incandescent. Assume also that the CF lasts for 9,000 hours, versus the incandescent's 1,000 hours. Finally, assume electricity costs 8 cents per kilowatt hour (kwh), roughly the national average.

Let's run the numbers. (We've rounded off to the nearest dollar.) To purchase and operate that one CF for 9,000 hours costs about $33 ($13 in electricity plus $20 for the bulb). To get 9,000 hours' worth of incandescents costs about $50 ($43 in electricity plus $7 for 9 bulbs). That's a savings of about $17 for the CF, despite its much higher purchase price.

Now, all this assumes that the CF actually does give off four times more light for the same energy, and that it really will be good for 9,000 hours—the equivalent of roughly a year of round-the-clock use. In reality, the manufacturers' ratings aren't always up to snuff. Andrew Rudin, a Philadelphia-based energy management specialist, measured the light output of several CFs and found manufacturers' claims wanting. For example, he measured a 27-watt Panasonic CF, supposed to be equivalent to a 100-watt incandescent. But the incandescent produced 43 percent more light. Similarly, says Rudin, claims of CF life expectancy may be exaggerated, but his evidence is more anecdotal than scientific.

Life-Cycle Thinking

Of course, any evaluation of light bulbs must take into account their cradle-to-grave impact. What about the pollution associated with

(continued on page 238)

LIGHT BULB BUYING CHART

COMPACT FLUORESCENT LIGHTING PRODUCTS	TOTAL POWER CONSUMPTION (LAMP & BALLAST)	APPROXIMATE INCANDESCENT EQUIVALENT	SAVINGS: AVOIDED UTILITY & LAMP COSTS* $	SAVINGS: AVOIDED CO_2 EMMISIONS	INDOOR: TABLE/FLOOR LAMP	INDOOR: BARE BULB SOCKET	INDOOR: OPEN PENDANT
STANDARD							
COMPACT 7-TWIN	10.8 W	40 W	$27	767 LBS	◒	◒	◒
COMPACT 9-TWIN	12.9 W	40+ W	$25	974 LBS	◒	◒	◒
COMPACT 9-QUAD	10.5 W	40+ W	$27	1,037 LBS	●	◒	◒
COMPACT 13-QUAD	15.3 W	40 W	$43	1,174 LBS	●	◒	◒
15-WATT CAPSULE	15 W	60 W	$39	1,063 LBS	◒	●	●
15-WATT GLOBE	15 W	60 W	$39	1,063 LBS	○	●	●
DULUX 11-WATT	11 W	40+ W	$27	761 LBS	●	◒	◒
DULUX 15-WATT	15 W	60 W	$43	1,182 LBS	●	◒	◒
DULUX 20-WATT	20 W	75 W	$57	1,444 LBS	◒	◒	◒
EARTHLIGHT	18 W	75 W	$59	1,497 LBS	●	●	●
18-WATT CAPSULE	18 W	75 W	$53	1,347 LBS	●	●	●
27-WATT E QUAD	27 W	100 W	$66	1,725 LBS	●	◒	◒
CANDLEFLAME	10.8 W	40 W	$27	767 LBS	○	○	○
SUN GLOBE	10.8 W	40 W	$27	767 LBS	○	●	●
HURRICANE	12.9 W	50 W	$35	974 LBS	○	○	○
FLOODLIGHTS							
PAR 38	10.8 W	60 W	$47	1,292 LBS	○	●	○
REFLECT-A-STAR	15.3 W	75 W	$61	1,549 LBS	○	●	○
SUPER NOVA	15 W	100 W	$84	2,333 LBS	○	●	○
GLOBE FLOOD	15 W	50 W	$44	828 LBS	○	●	○
DULUX FLOOD	11 W	60 W	$47	1,287 LBS	○	◒	○
EARTHLIGHT FLOOD	18 W	75 W	$59	1,497 LBS	○	●	○

* Over the lifetime of the compact fluorescent bulb at 8 cents per kWh electric rate, assuming residential application. Standard incandescent cost used: 75 cents. Standard floodlight cost used: $4.

FIXTURE APPLICATIONS					OUTDOOR FIXTURE APPLICATIONS			
CLOSED PENDANT	OPEN WALL	OPEN WALL	CLOSED WALL/CEILING	TRACK OR RECESSED CAN	PROTECTED FLOODLIGHT	POLE OR WALL MOUNTED	CEILING MOUNTED	TEMPERATURE RANGE
●	●	◒	●	○	○	●	●	-20 TO 130
●	●	◒	●	○	○	●	●	0 TO 130
●	●	◒	●	○	○	●	●	-20 TO 140
●	●	◒	●	○	○	◒	◒	0 TO 140
○	◒	◒	○	○	○	○	○	20 TO 95
○	◒	◒	○	○	○	○	○	20 TO 95
●	●	◒	●	○	○	●	●	-20 TO 140
●	●	◒	●	○	○	●	●	-20 TO 140
●	●	◒	●	○	○	●	●	-20 TO 140
●	◒	◒	◒	○	○	●	◒	-20 TO 140
○	◒	◒	○	○	○	○	○	-20 TO 95
○	●	◒	○	○	○	○	○	0 TO 95
○	○	●	○	○	○	●	○	-20 TO 120
○	○	●	○	○	○	●	○	-20 TO 120
○	○	●	○	○	○	●	○	0 TO 120
○	○	○	○	●	●	○	○	-20 TO 120
○	○	○	○	●	●	○	○	0 TO 120
○	○	○	○	●	◒	○	○	20 TO 140
○	○	○	○	●	◒	○	○	20 TO 95
○	○	○	○	●	◒	○	○	-20 TO 140
○	○	○	○	●	●	○	○	-20 TO 140

● BEST ◒ SUITABLE ○ NOT RECOMMENDED

(continued from page 235)

manufacturing and disposing of CFs? Here, CFs' advantage gets a bit murky. One reason is that some CFs use a radioactive material, as well as mercury, a toxic metal. The tiny amount of radioactive material—roughly the same amount used in smoke detectors—is used in the ballast, a device that provides a high-energy "kick" when the bulb is turned on. The radiation and mercury aren't thought to pose a problem—as long as the ballasts don't break. But they ultimately will, if only when disposed of in landfills.

Moreover, both mercury and radioactive materials place a heavy burden on the environment when they are mined and produced. According to Albert Donnay, a critic of the lighting industry and president of EcoWorks, which markets a long-life, energy-saving incandescent bulb, the radioactive materials used in CFs are "produced in special nuclear reactors dedicated to 'materials production,' " including plutonium for nuclear weapons.

Others disagree with Donnay over the extent of the radiation problem. Says Lindsay Audin, manager for energy conservation at Columbia University and a proponent of CFs: "It's not what goes into the manufacturing of the bulb, but its consumption of energy that's critical." Audin says the amount of mercury, lead, arsenic, and radioactive material released in the burning of coal is far, far greater than that used in CFs. "The same is true when you look at how much radioactive material goes up the stack of a nuclear plant—uncontrolled—to generate the electricity needed to power [incandescent] light bulbs. The differences are staggering. The amounts of electricity that can be saved by using CFs far outweighs the solid-waste problem, which should be addressed—for all lighting fixtures."

Note that not all CFs use radioactive materials, and these CFs seem destined to be phased out. The materials are used only in bulbs with *magnetic* ballasts, as opposed to newer *electronic* ballasts, which are radiation-free. The trend among all CF manufacturers is away from magnetic ballasts, so the radiation issue will eventually be moot.

Better yet, there are "component" CFs that allow you to retain the base when the bulb burns out, replacing only the bulb. Ballasts usually last for about four bulbs. These bulbs typically cost more than other CFs, but save money over time by aoiding the purchase of ballasts.

It can be difficult discerning which bulbs contain radioactive materials. You'll have to read the fine print carefully. GE's 15-watt

Care and Feeding of CFs

Here are some things to keep in mind about compact fluorescents:

- Make sure the CF is suitable for the kind of fixture in which you intend to use it. Some bulbs aren't suitable for enclosed fixtures, or outdoor use, or areas that are not well ventilated, or which experience extreme hot or cold.
- If the bulb is going to be used in a lamp, make sure the weight of the bulb won't make the fixture top-heavy. Also, make sure the bulb's size won't interfere with the lamp shade.
- Look for bulbs with electronic ballasts. These do not use radioactive materials. Ideally, look for a CF that features a permanent ballast and replaceable bulb.
- CFs are best suited for areas where lights will be left on for a while, such as porch lights, hallways, and other security lighting. In rooms where lights are rarely used, such as closets, stick to incandescents.
- As with any delicate electrical or electronic equipment, take care when handling CFs, especially in areas where they are prone to breakage.
- If you do break a CF, make sure to clean things up carefully. Try not to touch the components with your hands. If possible, air the room out for a while.

Compax, for example, states in the finest of type that "Starter bottle within lamp base contains 0.33 Micro-Ci Pm-147," a reference to Promethium-147, which, along with Krypton-85 (Kr-85), are the most commonly used materials.

The Limits of Compact Fluorescents

Another thing to keep in mind about CFs is that they are not suited for every use. As stated earlier, they can't be used with dimmers. They also may not be suitable for table lamps in which the lamp shade clamps onto the bulb. Another problem with using CFs with lamps is that most bulbs are considerably heavier than incandescents. This can make some fixtures top-heavy, increasing the likelihood that it will tip over (and break the bulb). CFs also aren't suited for lights that are turned on for only short periods. The reason: the electronic ballasts "burn out" a little every time they are used. A good rule of thumb is to consider CFs when the light will typically be on for several hours a day.

The chart on pages 236–237 show the various types of CFs , along with the amount of energy and pollution savings you can expect.

A Dim View

What about incandescents? They, too, have their place, and it makes sense to look for energy-efficient models. But be careful: Some companies have irresponsibly cloaked their products in green. Consider GE's "Energy Choice" bulbs. The 90-watt bulbs claim to use 10 percent less energy than a 100-watt bulb, but it's because they give off 10 percent less light. And be wary of the $2.40 energy savings boasted on the package. That's for 4 bulbs. Moreover, the light is harsher than GE's "soft white" bulbs—and more expensive, too.

EcoWorks bulbs boast 10 percent energy savings, too, but they also feature longer lives than most incandescents—up to 3,000 hours, four times that of GE's bulbs. **EcoWorks** (2326 Pickwick Rd., Baltimore, MD 21207; 301-448-3319) also boasts that it is "nuclear free" compared with GE, GTE Sylvania, and Philips, which are all involved in the production of nuclear weapons or nuclear power.

In the end, the impact of whatever bulb you buy is directly affected by how you use it. Simply turning lights off when not needed can go a long way toward saving resources and money. In other words, the most important energy-saving device in your home may be the switch on the wall.

Where to Find Compact Fluorescents

Here are names and addresses of compact fluorescent manufacturers. Contact these manufacturers for their latest product literature. See also the charts on pages 236–237 for other sources of CFs.

- **General Electric Co.,** Nela Park, Cleveland, OH 44112; 800-626-2000.
- **GTE Sylvania Lighting Div.,** 100 Endicott St., Danvers, MA 01923; 508-777-1900.
- **Osram,** 110 Bracken Rd., Montgomery, NY 12549; 800-846-2627.
- **Panasonic,** 1 Panasonic Way, Secaucus, NJ 07094; 201-348-5380.
- **Philips Lighting,** P.O. Box 6800, 200 Franklin Sq. Dr., Somerset, NJ 08875; 908-563-3000.

RECHARGABLE BATTERIES

The numbers alone are electrifying: Nearly 3 billion household batteries are purchased, consumed, and disposed of in the United States every year at a cost of more than $3 billion. They seem to power just about everything these days—toothbrushes, televisions, tools, toys, and countless other treasures of a gadget-mad world.

Despite the cute advertising images of drum-beating bunnies or swing-dancing toys, there is nothing inherently attractive about batteries, environmentally speaking. Each is a compact packet of poisons. When in use, the poisons are safely stored within a battery's shiny container. The problems come when the container is disposed of—usually in the trash. From there it goes to a landfill or incinerator.

That's where the trouble begins.

Batteries contain toxic metals, such as lead, mercury, and cadmium, all dangerous to human health. Mercury, for example, affects the central nervous system and kidneys. Cadmium can cause prostate cancer. Lead is linked to a variety of illnesses, from convulsions to kidney damage, and can affect brain development. When dumped in landfills, these metals can leach out of corroded batteries and into groundwater. If batteries are burned in incinerators, the metals can end up in ash (which is disposed of in landfills) or air emissions.

And yet we can scarcely do without batteries. For some people, they are practically a way of life.

The Powers That Be

The obvious alternative are rechargeables, which have grown considerably in sophistication, convenience, and availability in recent years. Once the domain of hobbyists, who purchased recharging systems from specialized catalogs, they are now being mass-marketed.

Nickel-cadmium batteries, or nicads, are the most widely used rechargeables. They come in the same sizes and voltages as disposable batteries, but they can be recharged as many as 1,000 times. Of course, the initial investment in rechargeables, including the plug-in recharger, is considerably more than their disposable counterparts. And the energy used for recharging costs money, too. But both costs will eventually pay for themselves many times over (see box on page 243).

Not that rechargeables are environmentally safe. They, too, contain toxins such as cadmium and lead, though some newer models use less-toxic nickel-metal hydride technology. These rechargeables use metallic alloys—including manganese, titanium, vanadium, and zirconium—to replace cadmium and lead, easing disposal problems.

The Limits of Rechargeables

Rechargeable batteries are not suitable for every application. For one thing, they self-discharge, meaning they lose their power just by sitting around. Hydrides, in fact, discharge twice as fast as nicads. With cordless appliances, this is overcome by keeping the device plugged into a charger. But what about the flashlight kept in your car's glove compartment? With a rechargeable, the power might not be there when you need it. (An alternative: Connect the rechargeable light to a photovoltaic solar panel on the dashboard.) So, for some situations, disposables may still be the best bet.

If you must buy alkalines, seek out those with little or no mercury. Recent years have seen a significant reduction in the use of mercury in alkalines—from about 1 percent to about 0.025 percent. (An exception is Ray-O-Vac, which contains 0.5 percent mercury.) In addition to the environmental benefits of reduced mercury, these batteries also minimize the danger to infants or pets from accidental ingestion of the batteries. Best of all, of course, are mercury-free batteries. Power Plus of America (1605 Lakes Pkwy., Lawrencelle, GA 30243; 404-339-1672) has introduced a mercury-free line: Powerplus Super Heavy Duty ES batteries. Power Plus claims they also contain 95 percent less cadmium than other heavy-duty batteries.

In the end, keep in mind that no battery will ever be good for the environment. But some do less harm than others. So the fewer batteries you use, the better for the earth.

Where to Get Rechargeables

Here are several sources for rechargeable battery systems. In general, charger prices range from \$20–\$30; batteries cost \$2.50 to \$5 each. Because "AA"-sized batteries are the most widely used, several companies also make charger models that accommodate only this size battery, charging them in as little as 1 hour.

• **Gates Energy Products** (P.O. Box 147114, Gainesville, FL 32604; 904-462-3911) makes the Millennium system, available in

Savings for You and the Earth

How much can you save by using rechargeables? It depends on how many batteries you use, of course. But here's a hypothetical example, based on an analysis done at Carnegie-Mellon University for the World Wildlife Fund:

Assume you are running a Walkman-like portable stereo for two hours a day for three years. The calculations assume the stereo uses four batteries. If they are alkaline batteries, they would last five hours; if replaced by nicad rechargeables, they would last only two hours. After three years of use, the stereo would have gone through 786 alkaline battery cells, costing $657. The four original rechargeable nicads would cost only $11. Adding the cost of the recharger ($10) and the electricity needed for 1,095 recharges ($5), the total cost of the rechargeables would still be only $26, or $631 less than the disposables.

Then, of course, there are the environmental savings. Those 876 alkalines would have amounted to roughly 54 pounds of trash, including 105 grams of mercury. The rechargeables, once eventually disposed of, weigh just over three ounces, including about 16 grams of cadmium.

hardware stores, Sears, and Seventh Generation (800-456-1177).

• **General Electric** (Nela Park, Cleveland, OH 44112; 800-626-2000) offers GE Rechargeable Batteries, available at Ace Hardware, Kmart, and Wal-Mart.

• **Harding Energy Systems** (826 Washington Ave., Grand Haven, MI 49417; 616-798-7033) makes the Green Battery, which contains no toxic metals, available primarily from stores specializing in eco-products and through catalogs, such as Real Goods (800-762-7325) and Jade Mountain (800-442-1972).

• **Radio Shack** (1500 One Tandy Ctr., Ft. Worth, TX 76102; 817-390-3011) sells its own brand of rechargeables through its own stores.

• **Sanyo Energy (USA) Corp.** (2001 Sanyo Ave., San Diego, CA 92173; 619-661-6620) offers the RechargAcell line of rechargeable batteries, available from Earth Tools (800-825-6460).

• **Sun Watt Corp.** (RFD Box 751, Addison, ME 04606; 207-497-2204) offers the F-2B solar battery charger. Using photovoltaics, it can charge up to six "AA" batteries in "less than one day," according to company literature, but "a full load of 'C' and 'D' cells will take two days or longer, depending on how sunny it is." Available from Jade Mountain (800-442-1972) and Co-op America (800-424-2667).

SAVING WATER AT HOME

We waste a lot of water, especially hot water. This wasn't always a problem, because there was plenty of water to go around. That's no longer the case, because of increasing populations, the greater likelihood of droughts resulting from global warming, more water pollution, and acid rain. All of this means that surpluses of water will increasingly be a thing of the past, and water shortages will become routine.

The typical American home uses almost 300 gallons of water a day. The fact is, you can save a great deal of that without making major sacrifices, and save money in the process in the form of lower sewer, water, and energy bills. If you have a septic tank—30 percent of the population does—conserving water will reduce the wear and tear on your system, and will require less energy from pumping well water. And then there are the tax dollars saved by not having to expand existing water treatment plants or to build new ones.

How do we waste water? Most is wasted at home. A steady faucet drip can waste 20 gallons of water a day. A leaking toilet can waste 200 gallons a day. At the water pressure found in most household plumbing systems, a $^{1}/_{32}$" leak in a faucet can waste up to 6,000 gallons a month.

Fortunately, there are a growing number of appliances and bathroom fixtures on the market that use considerably less water. In fact, it is likely that the water-saving features of many of these devices will be required in coming years.

Ten Steps for Saving Water

Here are ten simple steps you can take to minimize the amount of water you waste at home:

1. Test for a leaking toilet by adding food coloring to the tank (not the bowl; the tank is the covered portion of the toilet that contains the flushing mechanism). Without flushing, note if any color appears in the bowl after 30 minutes. If color appears, you have a leak.

2. Check your water meter while no water is being used. If the dial moves, you have a leak.

Flush with Success

Flush-N-Save Valve can turn most toilets into ultra-low water savers. The device, which retails for $14.95, fits all existing toilets of 3.5 gallons or larger, says the manufacturer, Derus Ecology Associates. The valve will cut water use to about 2.45 gallons per flush, a savings of 55 percent over the older 5.5-gallons-per-flush tanks in many homes. For more information, contact Trademark Sales & Marketing, 1920 American Ct., Neenah, WI 54956; 414-727-1818.

3. Turn off your hot water heater when going on a long trip.

4. When washing dishes, don't run water continuously in the sink. If you use a dishwasher, run it only when you have a full load and use the cycles with the fewest number of washes and rinses.

5. Add your garbage to the compost or trash instead of putting it down the disposal. Disposals not only use a great deal of water, they also add solids to an already overloaded sewer system.

6. Use a toilet dam to save water every time you flush. *Do not—repeat, do not—put a brick in your toilet tank to save water. The brick will eventually wear away and sediment may clog your pipes.* Most commercial dams are very inexpensive and can save up to 4 gallons per flush. Alternatively, fill gallon, half-gallon, or quart jugs with water and set them inside the tank. (This is an appropriate use for those plastic jugs you've already purchased but don't want to throw away.)

7. Install a low-flow shower head. It will reduce water use by up to a third without affecting the shower pattern. A normal shower uses about eight gallons of water per minute. Low-flow devices can cut the flow to 2 or 2 $^1/_2$ gallons per minute.

8. Install aerators on your faucets. These screw onto the faucet and add air to the water, giving a fuller flow with less water—about 40 to 60 percent less water. They also reduce splashing. Aerators are available at any hardware store for a dollar or two.

9. Water your lawn only during the coolest part of the day to avoid rapid evaporation.

10. Take showers instead of baths. A bath can use 30 to 50 gallons of water, compared to 20 gallons for a 10-minute shower with a low-flow shower head.

Pipe Cleaner

If you're leery of caustic drain cleaners and the effects they have on the environment and your plumbing, and haven't been satisfied using such home remedies as baking soda and vinegar, there's a new product that may be of interest. Topline Instant Drain Opener is a pressurized can that literally blows away blockages without using chemicals. One can works up to 20 times and is said to be safe for plastic and metal pipes as well as septic systems. For more information, contact Apex Distributing, 3171 Lenworth Dr., Unit #8, Mississauga, Ontario L4X 2G6; 416-624-0219.

Where to Get Water-Saving Devices. Many of the products described above can be found in local hardware stores or the houseware departments of department stores, as well as from these companies:

- **Co-op America** (2100 M St. NW, Washington, DC 20036; 202-872-5307, 800-424-2667) offers several water-saving devices in its catalog, including Europa Showerheads ($11.75 to $19) in a range of colors, from "china blue" to "cherry red" to "brite white."
- **Resources Conservation, Inc.** (P.O. Box 71, Greenwich, CT 06838; 203-964-0600, 800-243-2862) distributes a line of "Drought Buster" water-saving devices sold in hardware stores and home centers, including shower heads, faucet aerators, and toilet dams.
- **Seventh Generation** (49 Hercules Dr., Colchester, VT 05446; 800-456-1177) sells several water-saving devices, including a low-flow shower head ($11.95) and toilet dams ($5.95, three for $19.95). The "Water Saver Kit" contains two shower heads, two toilet dams, and two faucet aerators for $29.95.

Another helpful resource is *Turn Off the Tap*, by Randall Schultz, a simply written 60-page booklet on how to cut your household water use by up to 50 percent. It offers mostly the standard tips, but it's written and illustrated in a way the whole family can understand. Besides the inevitable tips, it also includes simple but thorough explanations of water problems and solutions. The book is $3.50 postpaid from EcoTime, 11024 Montgomery, Ste. 311, Albuquerque, NM 87111; 800-869-8520.

ECOLOGICAL HOUSE PAINT

Each year, Americans spend around $5 billion on house and building paint, more than a half-billion gallons' worth. Besides the stuff we buy, it is estimated that the typical American household has an additional four gallons or so of paint stashed away in a garage, closet, or basement. That's roughly another half-billion gallons.

How we buy, use, and dispose of that billion or so gallons of paint each year can have a big impact on the environment, and on you and your family's health.

What's Wrong With Paint?

A canful of paint is a canful of potential pollution. When it is used to paint inside, the result can be indoor pollution; when used outside, it can contribute to outdoor air pollution. This is true whether the paint in question is oil-based or water-based (latex). Indeed, latex paint is made up of molecules of oil paint suspended in water.

One problem has to do with something called volatile organic compounds (VOCs), which mix with nitrogen oxides (from motor vehicles, power plants, and refiners) in the atmosphere to form ground-level ozone, known also as photochemical smog. Volatile organic compounds are also produced by many other things, including aerosol propellants and some cleaning products. VOCs are of particular concern in southern California and other urban areas with high levels of smog. About 2 percent of all VOC emissions are attributable to paint. That makes paint a significant pollutant.

As recently as 40 years ago, most paints used in this country were oil-based, using solvents containing very high levels of VOCs. When water-based paints were introduced in the 1950s, they revolutionized the industry, offering products that were easier to use and clean up, with lower odors and fewer VOC emissions. However, oil-based paints are still preferred for many applications, especially enamels used for wood trim. Today, roughly a third of all paint sold is still oil-based.

The use of any paint can create some serious problems indoors. The Environmental Protection Agency has deemed paint to be the number-one indoor air pollution problem most Americans face.

Even in normal times, indoor pollution can be 100 times greater than that outside. After an application of oil-based paints, that figure can soar to 1,000 times higher.

VOCs aren't the only problem. Most paints are produced from nonrenewable raw materials such as petroleum and coal-tar oil. Making paint from these substances requires using toxic chemicals and tremendous amounts of energy. A host of other substances influence a paint's characteristics, acting as thickeners, drying agents, preservatives, and agents to fight fungi, bacteria, and insects. Some of these additives are highly toxic chemicals whose long-term effect on people and the environment is very difficult to measure.

The list of possible ingredients of paint is lengthy. Some products contain as many as 300 toxic substances, including arsenic, cadmium, chromium, lead, and mercury. A study conducted by Johns Hopkins University found 150 different carcinogens in household paint. And we're only talking about what happens when you use the paint, not what happens when you manufacture the stuff. It seems there are a score or more hazardous byproducts of paint making, including titanium dioxide, sulfuric acid, and chlorinated hydrocarbons.

Of course, when we paint, we rarely buy just that one product. Inevitably, we need paint removers or strippers, which are harsh chemicals used to prepare previously painted surfaces; and, of course, thinner, used to clean oil-based paints from brushes, hands, and other surfaces. Each of these contains additional VOCs and other potentially problematic ingredients.

Does all this make paint environmentally unacceptable? No. But it points up that care must be taken when selecting and using paint and related products.

Alternative Paints

It may be worth considering one of the many brands of alternative "natural" paint products. Most of these are manufactured with considerably fewer toxic ingredients made from such things as beeswax, plant waxes, linseed oil, and tree resins. Instead of petroleum-based solvents, they use plant oils; some brands, however, use small amounts of petrochemicals as well. Most derive their colors from earth- and mineral-based pigments.

Do these paints work as well as the well-established name brands? It depends who you ask. "If you are going to use one, make

Color Her Red

Buying "greener" paints isn't s always that simple, as evidenced by the recent case of Nancy Skinner, a Berkeley, CA, environmental activist and city council member. After Skinner and her friends applied the milk-based paint to her walls, it developed an odor so rank that Skinner and her daughter had to evacuate the premises and throw away most of their clothes and books, according to a front-page report in the *Oakland Tribune*. Apparently the paints, made by the German company Auro, lacked preservatives. Just like old milk, it went sour.

From there, the tale goes sour, too. To combat the smell, Skinner scrubbed the walls repeatedly—with bleach, then hydrogen peroxide. Then she repainted them with regular latex paint. Finally, she was forced to take the old walls down and put new ones up. That still didn't do the trick. At last report, Skinner and daughter were staying with friends and her lawyer was contemplating action.

All of which doesn't necessarily mean you should abandon the thought of using earth-friendly paints. The paints have been used safely for centuries, but under certain conditions, the milk protein can develop bacteria, particularly under damp conditions such as rainy or humid weather. Skinner was the victim of a bad batch of the stuff; the paint's distributor has admitted that the batch was spoiled and shouldn't have been sold. And experts say there are preservatives that can be added to milk-based paint to solve the problem.

sure that you are familiar with the product," says John Banta, owner of Baubiologie Hardware (P.O. Box 3217, Prescott, AZ 86302; 602-445-8225), which sells the AFM Safecoat line of paints. "For example, some people like to use trisodium phosphate to wash their walls prior to painting. With AFM paint, if it is applied on top of that it will turn everything into a sticky, gummy mess."

Getting a professional painter to try these alternative brands may be a challenge. "Some painters are downright hostile about the alternative paints," continues Banta. "They say, 'The paints don't work.' And yet I've got plenty of professionals who use them."

Another consideration is price. Alternative brands can be as much as double the price of good-quality name brands. For example, a gallon of AFM Safecoat gloss enamel costs about $29, compared to $15 to $20 for a standard latex paint.

Still another problem is that with alternative brands, your color selection is limited—to earth colors, of course. You will not find the hundreds of colors seen in the swatches available at your local hardware store. Livos PlantChemistry's water-based paint, for example, is available in 11 colors, mostly ochres, reds, browns, and umbers, plus green and aquamarine. AFM's Safecoat semi-gloss enamel comes only in white and bone, an off-white shade. Making bright reds, oranges, and yellows is not yet possible using plant-based materials.

Despite these shortcomings, the alternative paints are highly recommended for individuals with allergies or who are generally sensitive to chemicals. Indeed, the VOC fumes given off by drying paint can disrupt just about anyone's system. And the fumes don't necessarily end when the paint is dry to the touch; for the chemically sensitive, they can persist for months. With alternative paint brands, this problem is reduced considerably, sometimes entirely.

DEALING WITH LEFTOVERS

One problem with any paint is how to get rid of the leftovers. The easiest thing, of course, is to toss it in the trash or pour it down the drain or into the ground. Clearly, this is polluting and, in many areas, illegal. The paint will eventually drain into nearby streams or seep into groundwater supplies. Here are a few good rules to follow:

• First and foremost, buying wisely can eliminate disposal problems altogether. Take careful measurements of windows, doors, and walls. "Most people buy a gallon of paint even though they only need a quart," says Stephen Lester of the Citizen's Clearinghouse for Hazardous Waste. "Mostly, that decision is based on cost. People should start looking at what they need even if it costs them more. This is one time when buying in bulk isn't such a good idea."

• If you have more than half a gallon left over, try recycling it. Consider mixing several colors of similar paints together; you'll get a beige or gray color that may make a good primer. Don't mix oil paints with water paints, however. In Seattle, latex paint is collected by the city, mixed, and sold as "Seattle Beige" to local schools and hospitals.

To make recycling easier, separate oil-based from latex paint and interior from exterior paint. Make sure cans are properly labeled and less than two years old. You might want to donate leftovers to a local theater group, parks department, school, or organization, or take it to a community exchange. A growing number of towns have "drop and swaps" once or twice a year. If yours doesn't, consider organizing one.

• If you can't recycle the paint, take it to a local household hazardous

Good-bye, Old Paint

Some paint-related products can be more dangerous than the paint itself. That's certainly the case with paint removers and strippers, used to soften and dissolve an old finish so you can scrape it off before applying a fresh coat.

Some of the most effective strippers are also the most deadly. They contain methylene chloride, which can melt away almost any finish with ease and when breathed can also lead to anything from headaches to heart attacks. The government considers the chemical to be a human carcinogen. Still, products containing methylene chloride are still being sold, albeit with a warning label.

Less-harmful strippers use something called nonvolatile organic esters. They're much safer and give off hardly any odor, though they work somewhat more slowly. They're also more expensive—around $30 to $35 a gallon, versus $20 to $25 for brands containing methylene chloride. One of the lower-priced is 3M's "Safest Stripper." 3M claims it is so mild you needn't wear gloves, and it cleans up with water.

Another way to remove old paint is with a heat gun, which needs no chemicals at all. These devices, which resemble hair dryers, dispense heat in excess of 800°F, causing paint to blister; that makes it a cinch to scrape off. Heat guns are fast, but require care lest you burn yourself. You can also scorch some wood surfaces. Also, they don't work on clear finishes, such as varnishes and shellacs. Heat guns cost between $30 and $60 to purchase but use only about a quarter's worth of electricity to peel the paint from both sides of a door, considerably less than the equivalent amount of stripper would cost to cover the same surface.

waste collection site. In some areas, such wastes are collected at curbside once or twice a year. For more information about collections in your area, contact your local municipal or state government. Also check under "waste disposal" in the yellow pages to see if there is a pick-up service in your area.

Other Tips

Here are some other things to consider when painting to ensure maximum safety for both you and the environment:

- Never pour thinners or solvents down the drain or flush them down the toilet. Put them in tight-fitting jars or cans and have them picked up or delivered to a hazardous waste disposal site.
- Paint thinner can often be reused. In time, paint sludge settles on

the bottom of the container. Pour the clean solvent off the top and use. Stuff an absorbent material—kitty litter works well—into the can with the paint sludge. Let it dry completely before giving it to a hazardous waste handler.

• Unused paint thinner or stripper may be accepted by furniture refinish shops or paint contractors.

• If there is a small amount of latex paint left in a can, leave the can open to dry in a well-ventilated place. Once the paint is completely dry, it can be placed in the trash.

• Place all solvent covered rags and newspapers inside a metal container with a lid. Discard it as you would hazardous waste.

Where to Find "Green" Paints

Each of these companies makes claims about its products' low toxicity and natural or organic ingredients. Contact each for a current catalog and price list. To keep costs down, none of these companies routinely sends color samples, although they are available upon request.

• **AFM Enterprises, Inc.** (1140 Stacy Ct., Riverside, CA 92507; 714-781-6860) offers a wide range of nontoxic water-based paint and other home-repair products, including sealers, adhesives, joint compounds, grouts, strippers, waxes, and polishes.

• **Auro Organic Paints** (Auro-Sinan Co., P.O. Box 857, Davis, CA 95617; 916-753-3104) offers plant-based organic paints, waxes, lacquers, cleansers, and polishes manufactured in Germany and Austria.

• **The Glidden Company** (925 Euclid Ave., Cleveland, OH 44115; 800-221-4100) has introduced what it calls "the first conventional latex paint available in the U.S. which contains no petroleum-based solvents." Though not claimed to be natural or organic, the paint, called Glidden Spred 2000, does not emit VOCs. It is available in ready-mixed flat and semi-gloss interior latex, and is priced competitively with other so-called premium brands.

• **Livos PlantChemistry** (1365 Rufina Cir., Santa Fe, NM 87501; 505-438-3448) features a line of natural, nontoxic, and low-odor paints, oil finishes, varnishes, adhesives, waxes, and thinners.

• **Miller Paint** (317 SE Grand Ave., Portland, OR 97214; 503-233-4491) offers a line of "low-biocide," no-fungicide paints.

• **Murco Wall Products** (300 NE 21st St., Ft. Worth, TX 76106; 817-626-1987) offers nontoxic paints and a joint compound.

EARTH-FRIENDLY FURNITURE

What you're sitting on right now may be harmful to your health—or the health of the earth.

How can household furniture be an environmental hazard? The potential problems have to do with the way the furniture is made—the materials used and the manufacturing methods.

The problems with furniture fall into four categories:

- finishes that pollute the air during the manufacturing process;
- the use of wood grown or harvested in destructful ways;
- foam cushioning made with ozone-depleting ingredients; and
- synthetic and natural fabrics that can contain noxious chemicals.

Furniture makers are moving to green their industry, but they're moving slowly. Most efforts have centered around eliminating CFCs in foam seat cushioning, and in reducing fewer finishes that emit volatile organic compounds (VOCs).

But determining the "ingredients" of furniture isn't always easy. The typical table or chair isn't labeled with its components and their sources. Did that wood come from a sustainably grown source or from a threatened rain forest? Was the cushion made by a manufacturer that has eliminated CFCs, or by one that may not yet be so enlightened? There's often no way of knowing.

There are some labels to look for. **Ultracel**, for example, is a brand of furniture foam made without CFCs or methylene chloride; the latter has been listed as a "probable" cancer-causing substance by the federal government. Ultracel can be found in furniture made by the Horchow Collection, Miles Kimball, Pennsylvania House, Ethan Allen, and Drexel Heritage, as well as in furniture sold at J. C. Penney. It also comprises the foam cores in Sears' all-foam mattresses.

In Search of a Friendly Finish

The process of finishing furniture with a high-gloss brilliance can involve spraying as many as 20 coats of color, sealant, and lacquer. The process emits VOCs, which can contribute to air pollution. In highly polluted regions, furniture makers (as well as auto body shops and other industries) are being forced to reduce their VOC emissions.

Silk Purses From Sow's Ears

Building With Junk: A Guide to Home Building and Remodeling Using Recycled Material, a 162-page illustrated guide, offers a wealth of ideas and resources for putting to use some of the millions of dollars in building materials that author Jim Broadstreet maintains are thrown away every day.

For example, consider glass. "Old salvaged glass blocks are sometimes available and can be used in innumerable ways," writes Broadstreet, a semi-retired owner of an architectural mill-working business. "If they are not 'cleaned'—if the mortar is still on them—be sure you can get if off without breaking the blocks. Glass blocks can be used in walls between kitchen cabinets, for bathroom walls, etc. We have also seen articles on glass blocks for solar-gain walls, and, in one case, the blocks were drilled and filled with antifreeze water solution to act as heat storage."

Beyond this tidbit are sections covering concrete, masonry, wood, millwork and cabinets, insulation, roofing materials, metals, plastics, hardware, floor coverings, furniture, art, and more. Also included is helpful advice on obtaining and storing materials, and on building codes and regulations that relate to using reclaimed materials.

Building With Junk offers practical, money-saving, and ecologically sound advice—the kind we wish we'd see in more eco-books. The book is $22.95 ppd. from Loompanics Unlimited, P.O. Box 1197, Port Townsend, WA 98368; 206-385-5087.

And while we're on the subject, take a look at *Building With Nature Newsletter,* which covers some of the same ground. The bimonthly publication—written by Carol Venolia, architect and co-founder of the Natural Building Network and author of Healing Environments—focuses on "eco-healthy" design and construction. Six issues are $45 ($15 for those with low incomes), including a subscriber directory and reading list for those who pay full price, from Venolia, P.O. Box 369, Gualala, CA 95445; 707-884-4513.

Advances are being made with water-based finishes, which don't contain VOCs. But the process hasn't yet been perfected for all types of finishes, and there are some technical problems. Example: The water causes spray nozzles and tubing to rust.

One technique manufacturers are trying is to use one or two base coats of solvent-based finish, then applying additional water-based coats. Companies are also experimenting with alternatives to spraying. For example, the Knoll Group, a New York-based manufacturer, dips its furniture into vats of finish.

Michael McCahey, a former reporter for a furniture trade publication, offers this advice to earth-minded couch potatoes: Look at the product's label. "If there's nothing on the label about it being friendly, it probably has solvent-based finishes on it," he says. "Otherwise, they go out of their way to tell you."

An alternative is to buy unfinished furniture, then finish it yourself—using shellac, milk-based paint, oil, or another finish that doesn't contain solvents.

In Search of Good Wood

One problem with wood furniture is the source of the raw material. Much of the hardwood used for furniture comes from threatened rainforests. The process of building roads through the forests paves the way for further development: After the loggers leave, others use the roads to lead farther into the forest, where they slash and burn still more land to create grazing areas for cattle. Those same roads can also pave the way for diseases from foreigners to indigenous populations.

Rain forest activists advise consumers not to buy such tropical hardwoods as koa, zebrawood, teak, mahogany, and rosewood, or any other hardwoods unless you can be sure they are not endangered species and are logged using sustainable methods.

Two groups now certify lumber suppliers and furniture manufacturers to help consumers find sustainably grown wood sources. Scientific Certification Systems (SCS), formerly known as Green Cross, has certified some furniture from The Knoll Group. Another program, Smart Wood, is conducted by the Rainforest Alliance. (See page 259 for more on Smart Wood.)

There's Something in the Air

The question of furniture fabrics and particle board is one of degrees. People who are very sensitive to chemicals will be bothered by formaldehyde and other compounds in particleboard and fabrics; others may not notice any effects.

While formaldehyde is a suspected carcinogen, science has not yet confirmed its health effects, especially in indoor air situations, where many factors can contribute to a problem. Formaldehyde is widely used in glues for particleboard, plywood, and other laminates, and as a wrinkle-resistant treatment for fabrics, including clothing.

Formaldehyde and other compounds present in furniture tend to emit fumes, or outgas, over time. The amount given off is most concentrated when a product is brand new, and diminishes fairly quickly. The rate of penetration into the air falls off fairly rapidly, over a period of weeks, depending on humidity, heat, and ventilation.

There are ways to avoid formaldehyde or minimize exposure to it. Look for solid wood furniture, or laminated wood products that are sealed. Plywood is preferable to particleboard.

If you're ordering furniture to be delivered, ask for minimal packaging. That will give the chemicals a chance to offgas. Moreover, you'll cut packaging waste.

Bruce Tichenor, an environmental engineer with the Environmental Protection Agency's indoor air branch, cautions against having too much of what he calls fleecy material. "Carpet, drapery, upholstery materials, a lot of people say that these type of products increase the potential for biologics to form," he says, explaining that mold, dust mites, and other organisms can breed in those materials. "If you can't physically wash that type of material, avoid it," Tichenor says. "This is for people who have very, very sensitive allergies."

Another possible source of irritants is airborne substances that are absorbed by the materials. "Any fleecy materials will also act as sponges for other organic vapors," he says. "If you wax your floor, use a hairspray, paint rooms, or use any product that has a solvent in it, it turns out that those substances have a tendency to absorb vapor, which is then re-emitted at a later time."

Tichenor also suggests a small field test a sensitive person could try when considering a piece of furniture: "Stick your nose in it, smell it, and if it doesn't give you a reaction, it's okay."

Wouldn't that look rather silly on the showroom floor? Replies Tichenor: "You'd sit in a chair before you bought it, wouldn't you?"

Green Furniture Makers

- **Bales Furniture Manufacturing** (229 Utah Ave., S. San Francisco, CA 94080; 415-742-6210; free catalog) makes furniture of a lumber core covered with veneer and finished with water-based stains.
- **Ron Fisher Furniture** (406 S. First Ave., Marshalltown, IA 50158; 515-753-3414; $10 catalog) makes indoor pine furniture with water-based finishes. Some pieces available through Spiegel and Bloomingdale's.
- **Kingsley-Bate Ltd.** (5587-B Guinea Rd., Fairfax, VA 22032; 703-978-7200) makes outdoor teak and mahogany furniture sustain-

Furniture Buying Tips

- Avoid furniture made from hardwoods and temperate woods unless you're sure it comes from a sustainably managed source.
- Buy solid wood furniture. If it's made from pressed wood, make sure all edges are sealed to minimize releasing of gases in the particleboard.
- Seek furniture made with water-based finishes, or at least minimal amounts of solvents. Also preferable are finishes applied by dipping rather than spraying.
- Consider buying unfinished furniture and finishing it yourself, using milk-based paint, furniture oil, or a water-based finish.
- Don't buy furniture made with foam cushions and upholstery unless the product says it is made without CFCs.
- When ordering, ask for minimal packaging, so any compounds in the furniture have a chance to evaporate during shipment.
- Test your sensitivity to a new piece of furniture before buying by sniffing it up close.

ably harvested from plantations in Indonesia.

• **The Knoll Group** (655 Madison Ave., New York, NY 10021; 212-207-2200) makes office and residential furniture with low-VOC finishes, and a line of bentwood furniture from sustainably harvested maple, as certified by SCS.

• **Plow & Hearth** (301 Madison Rd., Orange, VA 22960; 800-627-1712) is a mail-order retailer of outdoor teak furniture.

• **Smith & Hawken** (25 Corte Madera Ave., Mill Valley, CA 94941; 415-383-4415) sells outdoor cedar furniture.

• **Summit Furniture** (5 Harris Ct., Bldg. 2, Monterey, CA 93940; 408-375-7811) makes indoor and outdoor furniture of Indonesian plantation-grown teak and mahogany, most of it unfinished; the indoor furniture is covered in cotton.

OTHER RESOURCES

• ***Interior Concerns Resource Guide*** and ***Interior Concerns newsletter*** (Interior Concerns Publications, P.O. Box 2386, Mill Valley, CA 94942; 415-389-8049) contain information for environmentally concerned architects and designers. The guide is $40 ppd.; the newsletter is $25 a year for six issues when ordered with the guide, or $35 separately.

GOOD WOOD

We don't think much about where our wood comes from. In fact, most of us view wood and wood products as ecologically desirable, especially when compared with plastics, which seems to be replacing wood increasingly in everything from dashboards to doll houses to desks.

But harvesting and processing wood can cause problems for the earth. Some woods, for example, comes from threatened tropical rainforests. Other wood comes from "old-growth" trees that are home to threatened species of animals. And some wood creates eco-threats when it is turned into other products.

All of this can make life confusing for those of us who try to spend our dollars in an environmentally responsible way. To start with, the path most wood takes from forest to market is a complex one at best. It is extremely difficult to trace the origins of many wood products. Distributors, salespeople, even manufacturers may be less than fully informed about the source of their raw materials.

It's difficult to overstate the impact of commercial logging operations on tropical rainforests. In parts of South America, Africa, and Southeast Asia, commercial loggers ravage pristine land, building roads and cutting trees that renders the land unusable for decades. Some 42 million acres of tropical rain forests are destroyed a year, according to the Worldwatch Institute. "At this rate, all our remaining tropical forests will be gone in approximately 20 years," says Pam Wellner of the Rainforest Action Network (RAN).

Some companies try to quiet critics by describing their logging operations as "selective," suggesting that they forego clearcutting operations by judiciously logging choice trees. Wellner points out that this "standard industry line" belies the fact that "for every one tree cut, 10 are destroyed by bulldozers and skidders" used to build roads and haul cut trees out of the woods. And eventually, she says, they must go ever deeper into the woods to find new trees to "select."

The damage done to these forests isn't simply a matter of despoiling untouched beauty, although that, too, is taking place. Tropical rainforests are just beginning to be seen as vast untapped resources for a wide range of needs. (See "Seven Environmental Problems You Can Do Something About" for more on tropical rain forests.)

Rain forests aren't the only source of environmentally undesir-

Board Feat

Not all hardwood is suspect. Much of it comes from domestic sources and from sustainably harvested tropical sources. But knowing the source of wood isn't easy. You usually have to ask. But as we said, getting reliable answers is difficult.

Help may arrive in the form of a new certification program operated by Rainforest Alliance. Its fledgling program will help identify woods "whose harvesting does not contribute to the destruction of rainforests" by certifying companies who use or sell wood or wood products that are *exclusively* "Smart Wood Products." The program will also certify sources of woods, including natural forests and plantations, large commercial projects and small community ones.

For more information on the Smart Wood program, contact the Rainforest Alliance, 270 Lafayette St., Ste., 512, New York, NY 10012; 212-941-1900.

able woods. Old-growth forests, primarily in the western U.S., are a key source of domestic hardwoods. Harvesting these woods affects not just endangered species (such as the notorious spotted owl) but also destroy efficient watersheds that protect against soil erosion throughout the West. According to *Architectural Record*, "The recent housing boom in Japan and the weak U.S. dollar have made Northwest timber more desirable, not only in Japan but throughout the world."

Which woods are affected? The most common are lauan (used chiefly for plywood), teak (salad bowls, kitchen utensils, furniture), and mahogany and rosewood (used in home and office furniture).

Hardwoods also show up in pencils (according to RAN, only Empire Berol USA now uses rainforest woods), disposable chopsticks (entire forests are reportedly cleared every year for these), picture frames, tool handles, and construction dowels.

BUILDING A BETTER FUTURE

The construction industry, one of the obvious large users of all woods, has been targeted by RAN and other groups and by the American

Pine and Dandy

The Rainforest Action Network (RAN) has published *The Wood Users Guide*, a comprehensive directory to timber and timber products to buy and avoid. The 68-page paperback describes "how the corrupt tropical timber industry is destroying rainforests," along with a list of the "bad guys" and product and company information. To obtain a copy, send $7.50 plus $2.50 shipping to RAN, 450 Sansome St., Ste. 700, San Francisco, CA 94111; 415-398-4404

Institute of Architects itself, which is developing guidelines on environmentally responsible building. For example, finding defect-free, straight-grained timber for some projects, builders must turn to old-growth forests, the principal source of such wood. Clearly, if you are building a house, addition, or any other project, it is important to have your contractor or architect specify woods that do not come from rainforests or old-growth sources.

Alas, that's about all you can do: specify, ask, and educate. Until a full-fledged labeling program is in place, your best bet is to ask questions, learn as much as you can, and share with others what you know about what makes wood good.

RESPONSIBLE TRAVEL

It's become axiomatic in our hype-driven society that if it works, oversell it. That certainly has been the case with most Green Consumer issues: If the public wants products to be biodegradable, recyclable, or "earth-friendly," then companies will deem their products to be so, even if such terms have little or no meaning.

Few green things are being oversold these days as much as eco-tourism. By whatever name—"alternative travel," "responsible travel," "softpath tourism," "low-impact travel," "nature travel," or even, as one group calls it, "environmentally sensitive, community-based travel to natural areas"—it is one of the fastest-growing niches in the travel business. For travelers, it also can be one of the most confusing marketplaces around.

What exactly is eco-tourism? As with most other "green" issues, the term means different things to different people. Indeed, it covers a wide range of travel, from simple backpacking treks to extensive—and expensive—jaunts to Third World countries. It can involve the most meager of accommodations or the most luxurious.

Whatever the mode or destination, the aim of eco-tourists is to tread lightly on the earth—not just physically, but also in relation to the economic, social, and cultural impact of travel on the people and places being visited. Eco-tourism seeks to enhance the economic development of lesser-developed areas, not just by pumping tour dollars into the local economy, but by helping locals to generate their own income as an incentive to preserve natural resources.

This is easier said than done. Even the most ecologically well-meaning tour operator can run up against a myriad of challenges and obstacles that require them to make compromises and accommodations. And not all eco-tour firms are well-meaning to begin with. Some operators are downright misleading, taking advantage of travelers' growing eco-concern. Says Virginia Hadsell, director of the Center for Responsible Tourism: "Too many operators are slapping the new term on the same old tours."

As the World Tours

Why should we worry about all this? First and foremost is the impact of travel on the ecology of the earth and its people. The fact is, much of our travel to foreign lands doesn't benefit many among the native population. Most tour operators are based outside the country to which travel is being booked. Even when tourist dollars do enter the local economy, they may do more harm than good. If all the money does is attempt to westernize the area, building roads and runways that will be used only by more tourists, it can lead to a spiral of dependency by the native population on foreign tourists. "It's not 'eco' if it's not supporting the local economy in a way that makes it more beneficial and prosperous for them to keep it natural than to develop it," says Shelly Attix, founder of Travel Links, an eco-tourism program offered by Co-op America. Natives have no incentive not to cut the rain forest down if most of the money spent leaves the country and doesn't benefit the natives, she says.

But money isn't the only issue. Through operators' poor planning and guides' ineptness, there have been many reports of disregard by eco-tourists for the land or its people. The horror stories are endless. Travelers tell of caravans of vehicles filled with camera-laden tourists encircling safari animals in the wild, jockeying to get good picture-taking angles. Others tell of trash heaps dumped in the wilderness, or of locals treated more like servants than hosts.

But many of the problems created by eco-tourists stem from sheer ignorance: They lack adequate knowledge of the culture or ecology of the places they visit. Some problems are caused by seemingly harmless goodwill gestures. "On many African safari tours, children congregate in places where up to 20 vans of tourists will pass each day," says Laurie Lubeck, a San Anselmo, CA, travel agent. "Each load of tourists think that they are being kind and

Eco-Tourists' Code of Ethics

- Travel in a spirit of humility and with a genuine desire to meet and talk with local people.
- Be aware of the feelings of the local people; prevent what might be offensive behavior. Photography, in particular, must respect people.
- Cultivate the habit of listening and observing, rather than merely hearing or seeing or knowing all the answers.
- Realize that other people may have concepts of time and have other thought patterns which are very different—not inferior, just different.
- Instead of seeing the beach paradise only, discover the richness of another culture and way of life.
- Get acquainted with local customs; respect them.
- Remember that you are only one among many visitors; don't expect special privileges.
- When bargaining for purchases, remember that the poorest merchant will give up a profit rather than give up his or her personal dignity.
- Make no promises to local people or to new friends that you cannot keep.
- Spend time each day reflecting on your experiences in order to deepen your understanding. What enriches you may be robbing others.
- If you want all the conveniences of home, why travel?

—Developed by the Christian Council of Asia

generous by giving these kids candy. In fact, the results are creating a mentality that they can get something for nothing—excess sugar in an already unbalanced diet. In the case of giving money, this results in kids earning more than their parents."

Rules of the Road

In light of such incidents, several groups have attempted to create codes of ethics for travelers. The National Audubon Society and the Christian Council of Asia have established criteria, and the travel industry is working on creating its own voluntary code.

Much of the codes involve common sense, the Golden Rule, and the motto, "Leave nothing but footprints, take nothing but photographs and memories." But it also involves some judgment calls. "If you are traveling thousands of miles by vehicle, you cannot carry all of your garbage in the back of your vehicle or you are going to have maggots and everything else traveling along with you," says Judy Wineland of Overseas Adventure Travel in Cambridge, Mass. Instead, she says, you'll need to check with park rangers or other locals to find a responsible means of trash disposal.

No matter how well behaved you may be, it is important to understand that just about everything you do has some impact on the area's natives, politics, economy, and natural resources. As with everything else, there are trade-offs. "It is important to create an authentic travel experience rather than build up a false fantasy of what the country's reality truly is, which contributes nothing to the host community," says Kurt Kutay of Wildland Journeys and founder of the Earth Preservation Trust.

Eco-tourism does offer benefits.For example, many one-time tourism opponents now see eco-tourism as a means to increase environmental activism. Once people see firsthand the beauty or even the hardship and destruction of an area, the theory goes, they may become emotionally involved and want to do something.

Seeing America First

Our references to Third World countries and indigenous populations isn't meant to imply that you must leave the country to be an eco-tourist. The United States has its share of wilderness, of course, much of it under the control of the **National Park Service**. (Contact them at P.O. Box 37127, Washington, DC 20013; 202-208-4747, for information about the park system or specific parks.) Most parks have environmental restrictions—limiting the number of horses, for example, or mandating trash disposal methods.

Native Americans represent an indigenous population that often suffers at the hands of tourists. "The Navajos want to do some interpretation and in some cases they also want to be left alone," says Jerry Mallett. "In their production of artifacts, they are trying to keep the quality up while keeping the price right. They are actually now taking stamped pots and painting them, instead of throwing the pots themselves. They sell them as Indian-painted, but the pots came from someplace else." Mallett believes we may be turning the tribes into pottery factories rather than learning from their values and cultures.

Picture This

The Center for Responsible Tourism is promoting a "Photographer's Code" that will help avoid causing problems when you tour foreign lands. Come to think of it, you would do well to apply the code everywhere. Among the suggestions:

- Leave your camera behind for a day. Try observing with the aim of memorizing scenes.
- Ask permission. In some places it is illegal as well as offensive to snap a person's photo without his or her permission.
- If you agree to send a print to someone, keep your word and do it. Get a complete address.
- Pay up cheerfully. Show no dismay if you are asked for payment in return for a photo.

Reprints of the entire "Photographer's Code" are available from the Center, P.O. Box 827, San Anselmo, CA 94979; 415-258-6594. To receive a complete publications list, send a stamped, self-addressed envelope.

Things to Find Out

Wherever you decide to travel, here are some things to find out about any tour operator or travel agent using environmental terms to describe their trips:

- Find out exactly what they are offering and what is their concern for the environment. Do they have standards or a code of ethics?
- Check credentials and expertise. Find out whether the operator actually has control over the operations in the country in which the trip is taking place. A lot of those offering trips are brokers.
- If the trip is to a wilderness area, find out the means of trash disposal. Simply throwing wastes into the wild is not acceptable; it can take years to degrade and may harm plants and animals.
- If animals are involved, ask how they are treated. Some tourists report ill treatment of pack animals.
- Ask whether they work with or are endorsed by any national environmental organizations. But that won't necessarily guarantee that trips will meet high standards.
- Most important of all, get references. Call or write someone who's been on a trip organized by the operator. A brief conversation may yield a wealth of information. If a tour operator won't provide

names and addresses of satisfied customers, be suspicious.

It would be nice if the word "eco-tourism" was redundant—that all travel was done with consideration for the physical and social environment. But that's hardly the case. In the meantime, eco-tourism is being seen increasingly as a responsible alternative to the theme park or bus tour, a way to truly enjoy the planet we are trying so hard to preserve.

ECO-TOUR ORGANIZATIONS

There is a growing number of local and national organizations offering information about ecological or low-impact travel.

- **Center for Responsible Tourism** (P.O. Box 827, San Anselmo, CA 94979; 415-258-6594) offers a wide range of information on eco-tourism. For more information send a self-addressed business-size envelope with 52 cents postage.
- **National Audubon Society** (700 Broadway, New York, NY 10003; 212-979-3000) offers a travel program stressing active participation in the preservation and conservation of natural resources. Land and sea trips cover areas ranging from Baja California to India.
- **Travel Link, Co-op America**, (14 Arrow St., Cambridge, MA 02138; 800-648-2667, 617-628-2667) is a full-service travel agency specializing in ecological and socially responsible travel. It donates 1 percent of its income to peace and environmental groups.
- **Wildlife Tourism Impact Project** (c/o Lori Lubeck, 9353 Duxbury Rd., Los Angeles, CA 90034; no phone) produces a safari sourcebook and a guide to deluxe, environmentally responsible adventures..
- ***Directory of Environmental Travel & Volunteer Activities***, by Dianne Brause. Send $10 to One World Family Travel, 81868 Lost Valley Education Center, Dexter, OR 97431; 503-937-3351.
- **The Ecotourism Society** (801 Devon Pl., Alexandria, VA 22314; 703-549-8979). A membership organization founded in 1990, it offers a newsletter and other resources on eco-tourism. Dues: $35 a year, $15 for students.

Other groups of note include **Earthwatch** (680 Mt. Auburn St.,

P.O. Box 403, Watertown, MA 02272; 617-926-8200), a nonprofit organization that funds scientific research expeditions. Travelers become members of research teams and share in both labor and costs of field research; part of the cost is tax deductible. Past programs include study of the environment of Australia's rain forest canopy, anthropologic digs that uncovered 4,000-year-old remains in Majorca, a study of spinner dolphins in Moorea, and observation of wildlife in the Blue Ridge Mountains of Virginia. Another organization, **Amazonia Expeditions** (18328 Gulf Blvd., Indian Shores, FL 34635; 904-332-4051), organizes trips that help support conservation efforts in Peruvian jungles.

Another group, the **Earth Preservation Fund** (EPF), is a nonprofit organization that sponsors such projects as reforestation in the Himalayas and cleanup hikes along the Inca Trail. EPF has information about responsible tourism and can arrange travel through its sister organization, **Wildland Adventures** (3516 NE 155th St., Seattle, WA 98155; 206-365-0686, 800-345-4453). You may wish to contact them about future programs and trips.

Travel Operators and Clubs

Here is a list of some of the organizations dedicated to green travel and a few adventure travel operators that specialize in nature tours:

- **Above the Clouds Trekking** (P.O. Box 389, Worcester, MA 01602; 508-799-4499) features in-depth cultural experiences, including trekking in the Himalayas, Andes, and Europe.
- **Adventure Center** (1311 63rd St., Ste. 200, Emeryville, CA 94608; 510-654-1879) offers safaris, cycling, sailing, and cultural interaction in more than 80 countries.
- **Alaska Wildland Adventures** (P.O. Box 389 Girdwood, AK 99587; 907-783-2928, within Alaska 800-478-4100) features a variety of safaris to Alaska with optional tent/cabin accomodations. Also offers King Salmon fishing program at their Kenai River Sport Fishing Lodge, an Executive Safari, accommodating small groups, a "Senior" Safari and Alaska Wildland Expeditions.
- **Biological Journeys** (1696 Ocean Dr., McKinleyville, CA 95521; 707-839-0178) features natural history marine wilderness tours in Baja California, Peru, Australia, New Zealand, and Alaska.
- **Backroads** (1516 5th St., Berkeley, CA 94710-1740; 510-527-

1555) offers deluxe bicycling, walking and cross-country skiing vacations throughout North America, Europe, Asia, the Pacific and the Carribbean. Guests travel from one place to another under their own power, either on foot, bicycles or skiis. A Backroads van supports the trip, carrying luggage, supplies and the occasional weary guest.

• **Caligo Ventures Inc.** (156 Bedford Rd., Armonk, New York 10504; 800-426-7781) arranges tours to Belize, Costa Rica, Panama, Venezuela, Trinidad and Tobago.

• **Cheeseman's Ecology Safaris** (20800 Kittredge Rd., Saratoga, CA 95070; 408-741-5330) features ecology safaris to East Africa, the Seychelles, Australia, New Guinea, South America, and Central America.

• **International Expeditions Inc.** (One Environs Park, Helena, AL 35080; 205-428-1700) offers eco-travel to a number of destinations, including Africa, Alaska, the Americas, Antarctica, Australia, Asia, the Chilean Fjords, Hawaii, India, Jamaica, New Guinea, New Zealand, and Panama. Also features International Rainforest Workshops to introduce travelers to rain-forested areas of the Earth.

• **Journeys** (4011 Jackson Rd., Ann Arbor, MI 48103; 800-255-8735, 313-665-4407) offers nature and culture-oriented travel to Africa, Egypt, Turkey, Norway, the Americas (including Hawaii), China, Tibet, Japan, Australia and New Zealand. Journeys has been helping local communities protect their natural and cultural heritage for 15 years.

• **Nature Expeditions International** (474 Willamette, P.O. Box 11496, Eugene, OR 97440; 503-484-6592) specializes in worldwide natural history and wildlife expeditions, focusing on photography and anthropology.

• **Safaricentre International** (3201 N. Sepulveda Blvd., Manhattan Beach, CA 90266; 310-546-4411) features wildlife and ecology safari tours throughout Asia, Africa, and South America.

• **Tamu Safaris** (P.O. Box 247, West Chesterfield, NH 03466; 800-766-9199) focus is on private safaris and adventure travel to six African countries: Kenya, Tanzania, Seychelles, Madagascar, Botswana and Zimbabwe, while ensuring that the environment is safeguarded. Revenue from the trips go to local people and local conservation efforts.

• **Tread Lightly Limited** (One Titus Rd., Washington Depot,

CT 06794; 203-868-1710) a full-service agency, dedicated to conservation and protection of ecosystem. Offers low-impact trips to natural areas as well and more traditional travel arrangements for leisure and corporate clients. Trip destinations include Belize, Bali and Borneo, Brazil, Chile, Mexico and the Yucatan Peninsula, Mongolia, Siberia and Newfoundland.

• **Victor Emanuel Nature Tours** (P.O. Box 33008, Austin, TX 78764; 512-328-5221) offers birding and natural history tours worldwide with a commitment to conservation.

• **Voyagers International** (P.O. Box 915, Ithaca, NY 14851; 607-257-3091) offers nature photography throughout Africa, Australia, Asia, and Antarctica.

• **Wilderness Travel** (801 Allston Way, Berkeley, CA 94710; 415-548-0420, 800-247-6700) offers wildlife- and nature-oriented trips as an active learning experience to increase awareness of preservation worldwide. Areas of interest include the Himalyas, South America, East Africa, and Europe.

• **Zegrahm Expeditions** (91414 Dexter Ave. N. #327, Seattle, WA 98109; 206-285-4000) features ecologically sensitive travel to Africa, Australian Outback, New Zealand, Asia and Pacific Islands, Amazon, Costa Rica, Sea of Cortez, Galapagos, Alaska, British Columbia, Maritime Provinces, Greenland, Canadian Arctic, NW and NE Passages, Trans Polar Bridge, Antarctica, Falkland Islands, and South Georgia Islands.

Educational Organizations and Volunteer Programs

In addition to the offerings through environmental organizations and travel outfitters, several educational centers provide natural history tours throughout the United States. Travelers can obtain credit for volunteering or participating in environmental programs offered by some of these programs.

• **Caretta Research Project—Savannah Science Museum** (4405 Paulsen St., Box B, Savannah, GA 31405; 912-355-6705) offers trips in which volunteers spend one week assisting project biologists monitor the nesting of endangered loggerhead sea turtles at the Wassaw National Wildlife Refuge and Georgia Barrier Island near Savannah. This is a volunteer conservation and research project rather than a strictly recreational travel opportunity.

• **National Audubon Society Expedition Institute** (P.O. Box 170, Readfield, ME 04355; 207-685-3111), in conjunction with Lesley College, offers unique courses leading to B.S. and M.S. degrees in environmental education. The program combines classroom study with one to two years of expedition courses. The program also offers a high school course that allows students to fulfill one year of accredited study.

• **Recursos** (826 Camino de Monte Rey, Suite A-3, Santa Fe, NM 87501; 505-982-9301) features natural history study trips of the delicately balanced desert ecologies of the Southwest. Travelers learn about such issues as desertification, decreasing plant and animal species, climatic changes, and the greenhouse effect.

• **Society for Ecological Restoration** (University of Wisconsin, Madison Arboretum, 1207 Seminole Highway, Madison, WI 53711; 608-262-9547) was established to promote research into the restoration, creation, and management of biotic communities. Participants provide hands-on help with ecological restoration projects in such places as Yellowstone National Park, Rocky Mountain National Park, and Afton State Park on the Saint Croix River in Minnesota.

• **Smithsonian National Associate Program** (Smithsonian Institution, 1100 Jefferson Dr. SW, Rm. 3045, Washington, DC 20560; 202-357-4700) conducts natural history study tours in a variety of locations. Foreign trips include China, India, Mongolia, and the Himalayas. Domestic destinations include Death Valley, Okefenokee Swamp, Sanibel Island, Alaska, and Appalachia.

• **Student Conservation Association** (P.O. Box 550, Charlestown, NH 03603; 603-826-4301) is a nonprofit educational organization providing students and others the opportunity to volunteer with resource management agencies maintaining national parks, forests, and public lands. A wide range of opportunities is available, with travel expenses and lodging provided.

• **University Research Expeditions Program** (University of California, 2223 Fulton, Berkeley, CA 94720; 510-642-6586) offers research expeditions emphasizing research that can be applied to improving people's lives and preserving the earth's resources. Travelers become part of a team and study such diverse subjects such as animal behavior, anthropology, sociology, archaeology, botany, and ecology in places such as Kenya, Indonesia, the United States, and the Fiji Islands.

Adventures for Special People

Environmental Traveling Companions (Fort Mason Ctr., Landmark Building C, San Francisco, CA 94123; 415-474-7662) specializes in wilderness adventures for people with special needs, including disadvantaged youth and people of all ages who are visually or hearing impaired, or physically, emotionally, or developmentally disabled. Trips span a wide range of activities, including sea kayaking, Nordic skiing, and white-water rafting. The group can provide school trips customized to meet the needs of teachers and students. Other individuals can participate in these trips by volunteering as guides.

Guest Houses and Resorts

• **Chanchich Lodge** (P.O. Box 37, Beliz City, Beliz; 800-343-8009 in the U.S., 501-(2)-75634 locally) offers a lodge and cabins in a tropical wilderness setting. Located in an ancient Mayan plaza, it features trails, wildlife, and Mayan ruins.

• **Lost Valley Center** (81868 Los Valley Lane, Dexter, OR 97431; 503-937-3351) is a family retreat and conference center located at the edge of the Willamett Valley near 200 square miles of national forests. The center provides campsites, cabins, and some dorm-style bunk units in a wilderness setting with many hiking trails and creeks and rivers nearby.

• **Maho Bay Camps, Inc.** (17 E. 73rd St., New York, NY 10021; 212-472-9453, 800-392-9004) offers environmentally planned camping on Saint John, U.S. Virgin Islands. The organization is also in the process of building a full-fledged resort area on Saint John, with flora and fauna restored to the pre-Columbus era, cabañas built off the ground to preserve tortoise habitat, and an emphasis on native wildlife.

Closer to Home

Many communities are involved in preserving the local environment. Indeed, there may be no better place to learn about the natural environment than a local zoo. A growing number of zoos are developing environments beyond cement cages, emphasizing animals' natural habitats. Contact the **American Association of Zoological Parks and Aquariums** (Oglebay Park, Wheeling, WV

26003; 304-242-2160) for information about zoological parks and aquariums in your area. For more information about zoos, endangered species, and animal protection, consult *The Nature Catalog* (Vintage Books, 1991).

There are many other ways to participate in environmental travel locally. Contact local environmental groups or local travel agencies for referrals to travel outfitters and other individuals in your area involved in responsible and environmental tourism. Below is a list of national organizations and government agencies that can provide information about local hiking, national parks, forests, seashores, and historical sites.

- **American Hiking Society** (P.O. Box 20160, Washington, DC 20041-2160; 703-385-3252) encourages hikers to build and maintain trails. Their "Volunteer Vacations" organize crews of volunteers to work on trail construction and maintenance for two-week periods on public lands across the U.S.
- **U.S. Forest Service** (Office of Public Affairs, Rm. 3008, South Bldg., Washington, DC 20250; 202-720-4623) has maps and information available for each of its 122 national forests.
- **National Park Service** (Office of Public Inquiry, P.O. Box 37127, Washington, DC 20013; 202-208-4747) supplies free information and maps for its national parks, forests, seashores, and historic sites.
- **National Parks and Conservation Association** (1015 31st St. NW, Washington, DC 20007; 202-223-6722) focuses on defending, promoting, and improving the national parks system and can provide information about specific sites and programs.
- **Rails-to-Trails Conservancy** (1400 16th St. NW, Washington, DC 20036; 202-797-5400) is dedicated to converting abandoned railroad corridors into public trails. There are more than 130 trails around the country for bikers, hikers, walkers, joggers, horseback riders, and even cross-country skiers. Routes include Ohio's Little Miami Scenic Trail, the Washington and Old Dominion Railroad Regional Park in Virginia, Iowa's Heritage Trail, and Florida's Tallahassee-Saint Mark's Trail.

TOYS AND GIFTS

Grown-ups love to rail on about today's toys and games—about how they've become exploitative, overpriced, and disposable, among other things. That certainly wasn't the case when *we* were young, right?

Perhaps. But more to the point is a dilemma some parents face every holiday season: How do you instill a sense of good, green values in a era when rampant consumption is among the first lessons children learn?

The best solution would be to spurn store-bought gifts altogether—to return to a simpler time when gifts were handmade, to be passed on to younger siblings, if not the next generation. And that is a worthy goal—if you've got the time, talent, and inclination. Unfortunately, most of us don't.

Which leads us to another option: a gift that will teach a child something about the earth and its environment.

The number of eco-oriented toys has grown dramatically in the last few years, and we've been continually impressed with their ingenuity and sophistication. The earliest generation of home-grown (and often amateurish) products has given way to more professional (albeit still sometimes home-grown) products.

Choosing Good, Green Gifts

That doesn't mean that choosing which, if any, eco-toy to buy is easy. Sometimes, a toy that is blatantly educational will be among the first to be put away and never used. "I think it's an unusual kid who gets excited by educational toys," says Alan Newman, former president of Seventh Generation, a mail-order company specializing in environmental products. "We're learning how to be very subtle, because our kids just don't have a lot of interest in mixing learning with fun."

Another challenge is to find toys that send the right messages. "I am very careful to select games that are very positive and uplifting, instead of those that are really portraying doom and gloom," says Donna Baker Schwenk, educational marketing manager for the National Wildlife Federation. She says the toys she chooses to sell are those that emphasize cooperation over competition.

One problem to watch out for, says Schwenk, are products that are complicated to use. The environment, after all, is hardly a simple matter. Explaining something as elemental as the ozone layer can involve rather complex concepts. "You learn through fun," she says. "If the rules and regulations and procedures for play are too extensive, then everyone is going to walk away from it." Schwenk's test is whether "a parent can pick it up and feel comfortable with it." Still another problem are games that are educational for the sake of learning, but with no clear goals. The same goes for toys and books that seem only to teach or preach, but don't leave kids with specific actions to take. Kids can get frustrated and cynical when they don't know what to do to solve a problem. The more positive actions they can take, the better.

Finally, keep in mind that some of the better toys and games aren't obviously environmentally oriented, but can still help instill an appreciation for nature and the earth. "I think they like the soil-testing-kit stuff, where they can get out there and explore their world and have the tools to do it with," says Jennifer Kaiser of The Nature Company, a mail-order and retail store chain. Among the products she recommends is the Ultimate Bubble Kit, which she says is pure fun while simultaneously teaching kids about physical properties. When it comes to nature, she says, "first you learn to love it, and then you'll want to protect it."

A GRAB-BAG OF ECO TOYS

Here is a sampling of environmentally oriented toys we've looked at for this holiday season, divided by age group. Note that many of the companies selling these items also have catalogs full of other toys and games. You are encouraged to contact them for more information.

Toddlers Through Age 7

• **Heart of the Dolphins** (356A Sebago Lake Rd., Gorham, ME 04038; 207-892-7367) sells Planet Pals. These double as musical toothbrush timers and crayon holders, designed to appeal to both sides of the brain. Made of domestic hardwoods, colorfully painted with nontoxic paints ($12 to $16).

• **Whale Gifts—Center for Marine Conservation** (P.O. Box 810, Old Saybrook, CT 06475; 800-227-1929) publishes *A Is for Animals* and *An Animal Shufflebook*, two clever activity books about animals. The first ($15.95) is a pop-up book for learning the ABCs; each letter has its own animal pop-up. The second ($14.95) is a set of bright cards with animals on one side, action words on the other. Each time you shuffle and lay the cards out, you get a new story.

• **Do Dreams Music Company** (P.O. Box 5623, Takoma Park, MD 20912; 800-4-Billy-B) sells "Billy B." tapes with upbeat melodies and lyrics. Tapes available are "Recycle Mania" ("So much trash, so much waste, filling up the land, taking up space; so much of it could be reused but that depends on me and you; Now there's just one question we all have to ask; If we can put a man on the moon, why can't we handle our trash?") and "Billy B. Sings About Trees," featuring lively songs like "My Roots Run Deep," "The Rock and Roll of Photosynthesis," and "Life of the Dead Tree." Newest releases include "Romp in the Swamp," rock to reggae songs about the wetlands and their inhabitants.

• **The Natural Choice** (1365 Rufina Cir., Santa Fe, NM 87501; 800-621-2591) offers Salis-Finger Paints, nontoxic with fragrance and colors derived from earthen and mineral pigments and all natural binders. Set of three (red, blue, and yellow), $9.95. Also available: Gellos-Wax Crayons, made from natural waxes and earth pigments. Set of 7 colors, $5.45.

Ages 8 and Up

• **Animal Town** (P.O. Box 485, Healdsburg, CA 05448; 800-445-8642) produces "Save the Whales," a game in which players work cooperatively to "save" 8 great whales, fighting against oil spills, radioactive waste, and other forces ($34). Also available: "Greenhouse Game" ($25), in which players move room to room through a house to learn unsafe household conditions.

• **The Daily Planet** (P.O. Box 1313, New York, NY 10013; 212-334-0006) offers Paper by Kids, a papermaking kit that includes everything you need (except a coffee can). Includes a hardbound book with easy-to-follow instructions ($24).

• **EarthAlert** (P.O. Box 20790, Seattle, WA 98102; 206-324-2362) is creator of "EarthAlert," a game that teaches environmental activism. Players compete in trivia and role playing while participating in "earth-saving" activities: turning off a light in a room, writing to a congressperson, and so on. Printed on recycled paper, with a portion of proceeds going to 5 environmental groups ($29.95).

• **Educational Insights** (19560 S. Rancy Way, Dominguez Hills, CA 90220; 800-933-3277) sells "Exploring Ecology" and "'Eco-Detective," hands-on kits containing 35 experiments each. "Exploring Ecology" ($34.95) shows kids how environmental changes take place. "Eco-Detective" ($12.95) explores the interrelationship of living things and their surroundings.

• **Hubbard Scientific** (P.O. Box 760, Chippewa Falls, WI 54729; 800-323-8368) sells Sunpower House, a working model of a passive solar house. Kids can conduct experiments that can help them understand solar energy and the greenhouse effect. Works with direct sunlight or a sun simulator lamp ($32.40). Also available: a set of 3 solar time pieces that use the sun to figure time and date ($21.50).

• **National Geographic Society** (P.O. Box 2118, Washington, DC 20013; 800-638-4077) features *Project Zoo*, a computer game about zoo animals. Kids (ages 8 to 12) design a zoo where animals, staff, and visitors will all be happy and healthy. For Apple II computers only ($49.95). Also available: a wide range of illustrated books for all ages.

• **The Nature Company** (headquarters: 750 Hearst Ave.,

Teaching Kids About the Earth

Keep in mind that you can teach kids about the environment perfectly well without any of these products. All it takes is a little time and patience. Here are a few tips:

- Admit what you don't know. It's okay to shrug your shoulders. These are complex subjects. It might provide a chance for you to find out together.
- Use simple explanations. Try to find examples of things kids can relate to.
- Give positive steps to take. Kids get frustrated if they can't solve a problem. Show them what they can do in their lives. Praise or reward their green actions much as you might other responsible behavior.
- Avoid fanaticism. Kids have a hard time knowing where to draw the line, so they're likely to go overboard—not letting you buy anything made of plastic, for example. Try to give kids choices and flexibility.

Berkeley, CA 94710; 510-644-1337, 800-227-1114; 90 stores nationwide) offers a wide range of products, from games to books, on a variety of subjects. Fun and educational items include the "Ultimate Bubble Kit" (age 8 and up) which explains and demonstrates the physics of bubbles and the "Bug House Kit," (age 5 and up) a collection of things for a young bug collector to use when searching for and capturing bugs.

• **Real Goods** (966 Mazzoni St., Ukiah, CA 95482; 800-762-7325) sells solar-powered wooden toys, assembled from simple kits in about an hour, that demonstrate the potential of the sun. Three models are available: helicopter, windmill, and airplane ($18 each).

GREEN GIFTS FOR EVERYONE

Clothing

• **Co-op America** (2100 M St. NW, Ste. 403, Washington, DC 20037; 202-872-5307, 800-424-2667) works with new vision companies offering a selection of clothing made by Mayan people in Guatemala and disadvantaged women and handicapped persons living in India. Catalog includes selection of cotton T-shirts that promote positive change through inspirational messages.

• **Deva Lifeware** (P.O. Box C, 303 East Main St., Burkittsville, MD 21718-0438; 301-663-4900, 800-222-8024) mail-order catalogue operation offering quality, handcrafted natural clothing for men and women. Clothing is sewn by a network of home-based stitchers who create each garment from start to finish with personalized tags with their names sewn in.

• **Environmental Awareness Products** (3600 Goodwin Rd., Dept. GRC, Ionia, MI 48846; 517-647-2535) carries clothing, jewelry, and gifts with environmental and cross-cultural designs. Also adding endangered species jigsaw puzzles, cookie cutters and various other wildlife gift items. Ten percent of proceeds go to various grassroots environmental organizations.

• **Jim Morris Environmental T-Shirts** (P.O. Box 831, Boulder, CO 80306; 303-444-6430) offers a wide variety of colorful T-shirts and sweatshirts with environmental and conservationist motifs. Environmental books from a variety of publishers are also available. At least 10 percent of profits are donated to environmental groups.

• **Marketplace: Handwork of India** (1461 Ashland Ave., Evanston, IL 60201; 708-328-4011; 800-726-8905, weekdays 9 a.m. to 4 p.m. Central Time), a nonprofit organization providing employment to 240 women and handicapped persons. Specializes in colorful cotton clothing in a variety of prints and international styles.

• **Seventh Generation** (49 Hercules Dr., Colchester, VT 05446; 800-456-1177) features clothing made from organic cotton and Fox Fibre, which grows in color so there is no need for bleaches and dyes. Green cotton tanks, polos, and cardigans come with tagua nut buttons which are grown in tropical rain forests. The Tagua Button Company is reviving the tagua button using sustainably harvested nuts and increasing the economic value of the rain forests. A portion of their proceeds goes to the Rainforest Alliance.

• **Wearable Arts** (26 Medway, Unit 12, San Rafael, CA 94901; 800-635-2781) produces a line of full-color, screen-printed, 100-percent cotton T-shirts. Some designs are available on unbleached, undyed, natural T-shirts with environmental messages. Designs include tropical birds, jungle scenes, coral reefs and the North American wolf and eagle. A percentage of the sales are given to the Earth Island Institute of San Francisco, CA.

Tunes for Tots

Music For Little People is an innovative mail-order firm specializing in kids' music, videos, instruments, and other items, many with environmental or socially responsible themes. A sampling from its latest catalog includes "Voices of the Rainforest," by Mickey Hart ($9.98 cassette, $16.98 CD); *The Man Who Planted Trees*, a book/album combination ($21.95); *The Kids Nature Book*, with 365 indoor and outdoor activities for ages 2-10 ($12.95); "Down the Drain," a 3-2-1 Contact video about water use and conservation for ages 5 and up ($15.98); and "The Rotten Truth," a Children's Television Workshop video about garbage, waste disposal, and recycling ($15.98). There is also a series of three audiocassettes performed by the Banana Slug String Band, featuring "Slugs at Sea," "Dirt Made My Lunch," and "Adventures on the Air Cycle" ($9.98 each, all three for $23.98).

There's more, including "Mother Earth," a cassette by Tom Chapin, a former host of TV's National Geographic Explorer, who sings about the environment and other topics ($9.98); and "Man of the Trees," a 25-minute video about Richard St. Barbe Baker, whose work to save forests on three continents is intended to show what one person can do to change the world ($19.98; $1 from each video is donated to Children for Old Growth).

For more information or to obtain a free catalog, contact Music for Little People, Box 1460, Redway, CA 95560; 800-346-4445.

Recycled Paper Products

• **Acorn Designs** (5066 Mott Evans Rd., Trumansburg, NY 14886; 607-387-3424) offers stationery, note pads, bookmarks, note cards, gift tags, and prints made with recycled paper.

• **Atlantic Recycled Paper Company** (P.O. Box 39179, Baltimore, MD 21212; 301-323-2676) offers recycled paper products with the highest amount of post-consumer waste possible, including unbleached toilet paper, paper towels, computer paper, envelopes, etc.

• **Conservatree** (10 Lombard St., Suite 250, San Francisco, CA 94111; 800, 522-9200, 415-433-1000) sells office and printing paper, including letterheads, envelopes (both plain and window), computer paper, copy paper, offset, coated papers, and newsprint, as well as text and cover papers appropriate for manuals, flyers, newsletters, and brochures. Conservatree will do an environmental impact analysis of office paper usage showing the positive environmental impacts of switching from virgin paper to recycled paper.

• **Co-Op America** (2100 M St. NW, Ste. 403, Washington, DC 20037; 202-872-5307, 800-424-2667) distributes recycled paper products, including facial paper, self-stick notes, kitchen towels, and a personal journal with 50-percent recycled paper. Also features 100-percent recycled, 20-percent post-consumer unbleached envelopes, "Preserve the Fragile Habitat" note cards and gift wrap from Earth Care Paper Company.

• **Earth Care Paper, Inc.** (P.O. Box 7070, Madison, WI 53704; 608-223-4000), offers a variety of paper products including stationery, writing pads, office paper, journals, gift wrap, and greeting cards. Earth Care's catalog contains a wealth of information on recycling and environmental concerns. Ten percent of profits is given to organizations working to solve environmental problems.

• **EJW Products** (P.O. Box 599, Quogue, NY 11959; 516-653-6437, 800-281-7137 in New York) sells recycled office paper for computers, copiers and fax machines, recycled file folders, adding machine tapes, printing paper, etc.

• **John Rossi Company** (180 S. Highland Ave., Ossining, NY 10562; 914-941-1752) produces "Revived Page" stationery, envelopes, hand-made writing pads and books, and paper place mats. The warm gray- and earth-toned paper is made from 100-percent recycled newspapers. Products are available in stationery stores, museum shops, and some mail-order companies.

• **Seventh Generation** (49 Hercules Dr., Colchester, VT 05446-1672; 800-456-1177, orders only) distributes a variety of products made from recycled paper including toilet paper and map stationery made from real maps from government surplus. Also see unbleached paper towels and napkins. Send $2 for catalog.

Solar-Powered Gadgets

• **Jade Mountain** (P.O. Box 4616, Boulder, CO 80306; 303-449-6601) offers various solar-powered gifts, including a "cool cap," a safari hat, a tea jar, flashlights, battery chargers and a solar-powered speed boat.

• **Real Goods** (966 Mazzoni St., Ukiah, CA 95482; 800-762-7325) offers a wide selection of solar-powered gadgets and toys including a solar-powered watch, a AM/FM radio and a "4-in-1 Solar Construction Kit," for children to build solar-powered windmills, airplanes, and helicopters.

• **Electric Engineering** (116 Forth St., Santa Rosa, CA 95401; 707-542-1990) sells a solar-powered sports radio with earphones. It takes three hours in the sun to generate enough power for four hours of playing time; it also runs on conventional AA batteries.

• **SunWatt Corporation** (RFD Box 751, Addison, ME 04606; 207-497-2204) offers solar battery chargers and rechargeable batteries. Plastic casing for the F-2B, which recharges AA, C or D cells are made from recycled plastic by the Penobscot Indians in Maine. Charger parts are made in-state using electricity from renewable resources and all parts are made in the United States.

Unusual Gifts

• **Basic Foundation** (P.O. Box 47012, Saint Petersburg, FL 33743; 813-526-9562) plants hardwood and fruit trees in endangered forests in Costa Rica (see "Garden and Pet Supplies" chapter for more information). A gift of $5 pays for a single tree; $250 for planting a hectare—containing approximately 1,000 trees.

• **National Wildlife Federation** (1400 16th St. NW, Washington, DC 20036; 800-432-6564) offers a host of nature-related gifts, including watches and EnvirOmints chocolates.

• **The Natural Choice** (1365 Rufina Cir., Santa Fe, NM 87501; 505-438-3448) offers clipboards, diaries, and photo albums made from recycled materials and covered with marbled paper. Also, hair combs made from sustainably harvested hardwoods.

• **Programme for Belize** (P.O. Box 1088, Vineyard Haven, MA 02568; 508-693-0856) offers an opportunity to become a shareholder in the future of Rio Bravo Conservation and Management Area and an investor in efforts to preserve biodiversity in Belize. Money from investors will provide operating funds for protection of the land and its resources. In addition to the regular "buy an acre" certificates at $50, donors can purchase $100 certificates indicating an "investment" in the future of Rio Bravo.

• **Tarlton Foundation** (50 Francisco St., Ste. 103, San Francisco, CA 94133; 415-433-3163) offers an Adopt-a-Whale program for $50 per year ($25 for youth groups), which is used to support education and research programs. Members receive a photograph of "their" whale mounted on an adoption certificate, along with biographical information and a 4-page whale fact sheet.

ALTERNATIVE TRADE ORGANIZATIONS

A growing number of Green Consumers are seeking out products made by people living in environmentally threatened regions of the world. These products range from baskets to clothing to pottery, as well as nuts and other products from rain forests.

The idea behind supporting these artisans is that by helping them support themselves through their crafts, we not only help raise their often meager standard of living, we also support their efforts to support themselves through traditional low-tech means. Poor citizens who aren't able to support themselves are more likely to allow their forests to be cut down and other natural resources to be exploited. In most cases, this brings only short-term gain to the native people, if it brings any gain at all. So, buying their crafts and other products is an environmentally responsible decision.

The bad news is that buying native crafts may not necessarily help the natives. By the time some products go from exporter to shipper to distributor to retailer, there is often little money left to pay the products' creators. According to Worldwatch Institute, a $14.95 woven basket from an outlet such as Pier One Imports might only yield $1 to the Philippines woman who spent a full day collecting, preparing, and weaving the grass, rattan, and bamboo that went into its manufacture. Of the remaining $14, about $3 goes to an exporter and another $9 may go to such expenses as import duties, distribution, shipping, handling, and advertising. So, Pier One may make only about $2 on the basket.

To help ensure that native craftspeople are paid a fair wage for their goods, a series of alternative trade organizations (ATOs) have emerged. ATOs aim to create fairer trade relationships by dealing directly with artisans and with small producer cooperatives that have demonstrated a commitment to social and economic justice. Some ATOs also provide these cooperatives with help in such areas as bookkeeping, shipping, and packaging.

Below is a listing of ATOs in the United States. Many publish catalogs; some sell exclusively to retailers. You are encouraged to contact each for their latest offerings and outlets.

- **Co-op America** (2100 M Street, NW, Washington, DC

20037; 202-872-5307, 800-424-2667) brings consumers together with alternative businesses. The retail catalog, published twice a year, includes jewelry, food, and household items produced by socially responsible businesses.

• **The Crafts Center** (1001 Connecticut Ave. NW, Washington, DC 20036; 202-728-9603) is an international resource center and network for artisans, providing technical assistance in product development and marketing. It publishes *Craft News* and an international directory.

• **Cultural Survival Enterprises** (53-A Church St., Cambridge, MA 02138; 617-495-2562) is a nonprofit human rights group working to "economically empower indigenous peoples and ethnic minorities." CSE works with citizens on four continents to sell their products, an extensive line of food and body-care items.

• **Equal Exchange** (101 Tosca Dr., Stoughton, MA 02072; 617-344-7227) imports and markets food from democratically organized, small-scale Third World farmers. It carries a full line of coffees and imports honey from co-ops in Mexico.

• **Guatemalan Refugee Crafts Project** (4315 Ave H, Austin, TX 78751; 512-323-6878) is a nonprofit self-help program run by Guatamalan refugees. The project is designed so that all members of two refugee communities in Chiapas, Mexico, are benefited through the production of high-quality, backstrap loom woven products. Products include small bags, wallets, belts, pillow covers, place mats and napkins, bedspreads, and hupils (traditional blouses).

• **Marketplace** (1461 Ashland Ave., Evanstown, IL 60201; 708-328-4011) works with women and handicapped persons in India, offering highly detailed, handmade dresses, tops, vests, jackets, pants, skirts, and scarves.

• **Mayan Hands** (992 Brookline Ave., Albany, NY 12203; 518-438-5636) works with Mayan artisans in Guatemala and Chiapas, Mexico, to design, develop, and market their products in the United States. Items carried include tablecloths, backpacks, jackets, throw pillows, bookmarks, and embroidered T-shirts.

• **Odara** (274 Taconic Rd., Greenwich, CT 06831; 203-869-8438) works directly with craftspeople in the Northeast and Amazon regions of Brazil to bring Brazilian folk art and culture to North America. Odara's most successful product is Rainforest Rubber Crafts—animals and figurines hand-molded from a sustainably extracted latex. Other products include handpainted rain sticks, rain-forest seed necklaces, traditional musical instruments, coconut crafts, painted ceramic figurines, and cotton rag rugs.

• **Pueblo to People** (2105 Silber Rd., Ste. 101, Houston, TX 77055; 713-956-1172) purchases from throughout Latin America. Products include handicrafts, clothing, cashews, and coffee. It works with cooperatives of the very poor that have organized themselves and are doing more than making products, such as providing educational or health services in the community.

• **SERRV** (500 Main St., Box 365, New Windsor, MD 21776; 800-423-0071) purchases and markets handicrafts of people in developing countries. It sells handicrafts from over 40 countries through two retail stores and by wholesale to more than 3,500 church and community groups.

• **Selfhelp Crafts of the World** (704 Main St., Akron, PA 17501; 717-859-4971), a project of Mennonite Central Committee, purchases from producers who pay their craftspeople a fair wage and those having a concern for improving the lives of their employees.

• **Silk for Life Project** (1300 S. Layton Blvd., Milwaukee, WI 53215; 414-384-2444) is an international crop substitution effort that aims to replace coca farming (the plants used to make cocaine) with silkworm cultivation. It guarantees the purchase of all silk cocoons and yarn produced by participants, and imports the yarn for use in workshops, where it employs women to make handwoven scarves. Products include hats, headbands, gloves, and scarves.

• **Thai International Display** (412 Zack St., #307, Tampa, FL 33602; 813-229-8292) is a crafts wholesaler and distributor that also works directly with primarily Asian craftspeople.

• **A Thread of Hope** (Box 1902, Eugene, OR 97440; 503-687-6865) links Mayan weavers with North American consumers, returning 70 cents of every dollar in sales to the artisans. The money is used to fund courses in literacy, numeracy, sewing, and leadership training.

• **Trade Wind** (P.O. Box 380, 156 Drake Ln., Summertown, TN 38483; 800-445-1991, 615-964-2334) links Native Americans and consumers. It runs a mail-order catalog and has just begun to offer wholesale orders to socially responsible stores and vendors.

• **Wicahpi Vision** (c/o Loretta Afraid of Bear Cook, 358 Bordeaux St., Chadron, NE; 308-432-2502) is a group of over 100 Lakota (Sioux) Native American artisans. Items featured include bead work, jewelry, and star quilts.

HOW TO GET INVOLVED

HOW TO GET INVOLVED

From letter-writing campaigns to company boycotts to street demonstrations, Americans have always found ways to get involved in the issues of the day. And when it comes to the environment, there is no shortage of issues on which you can get involved and make your voice heard.

The first step before getting involved is often contacting and supporting one or more of the organizations working on the topics about which you are concerned. (See page 306 for names, addresses, and brief descriptions of the key environmental groups. Many of these groups also have local chapters.) Most of these organizations can provide you with additional resources for understanding these issues, and may offer helpful suggestions on ways you can be effective in making change happen. While you usually need not be a member to obtain information, most memberships are relatively inexpensive—about $25—and besides being tax deductible, may entitle you to additional information or to discounts on literature.

Don't be discouraged by the fact that you "are only one person" or that your tiny local chapter of a big national environmental organization may not seem very powerful. Increasingly, much of the power of environmental change has come from concerned individuals and from local community groups taking action on issues close to home. As the number of such actions has grown, so, too, has the message to national and local government and to corporations: Americans are more concerned than ever about the state of the environment, and they intend to do something about it.

TAKING ON LOCAL POLLUTERS

One of the most effective ways individuals can get involved is to target local polluters, and to force action that will stop their poisoning of your community. Consider a few cases:

• In Lynchburg, TX, Sandra Mayeaux, a homemaker, spearheaded an effort to close two nearby toxic waste incinerators after she counted the number of her neighbors who had died of cancer. Operators of both incinerators—one of them just a half mile down the street from her house—were ultimately charged with a variety of permit violations in the dispersion of cancer-causing chemicals.

• In Berlin, NJ, citizens forced a local glass factory to adopt measures that reduced the chances of a toxic spill in the event of an accident.

• In Louisiana, a group of citizens organized the "Louisiana Great Toxic March" to bring attention to the high cancer rates in an area between New Orleans and Baton Rouge known as "Cancer Alley." After a rally at the state capitol, the marchers followed the Mississippi River for 130 miles to New Orleans, holding rallies, teach-ins, and press conferences at every stop.

• In Galveston County, Texas, Rita Carlson, whose group had only ten active members, organized a lawsuit on behalf of more than 1,000 people who required medical treatment after hydrochloric acid leaked from a nearby oil plant.

Unfortunately, most victories do not come quickly or easily. The wheels of justice seem to grind very slowly when it comes to environmental issues. One reason for this is that both local and national environmental laws are fairly weak, and enforcement of them is even weaker. Some companies have continued to pollute openly for years, knowing that it will take years—decades even—for government authorities to make them stop. By then, the small fines that will be levied against the polluters will be a drop in the bucket compared with what it would have cost the company to modernize its facilities or take some other action to stop dumping toxic wastes into the air and water. And when government agencies do respond they are often outgunned by the high-priced legal staffs of some of these companies, who are adept at finding loopholes in the law or otherwise stalling and confounding the authorities.

More effective action often requires filing a lawsuit. That may sound intimidating to most individuals and local environmental groups, but the time and expense may be worth the trouble. Lawsuits are time-consuming and expensive, to be sure, and are often settled out of court in a compromise that pleases neither party. But even a partial victory over a polluter can have a positive effect on stemming the problem.

One excellent resource, *Making Polluters Pay: A Citizen's Guide to Legal Action and Organizing* by Audrey Owens Moore, describes the nuts and bolts of the legal process in layman's terms. Designed as a step-by-step workbook for community groups and individuals, the book describes the legal, research, study, and medical efforts required to combat a local polluter, including how to gather information, the barriers toxic victims now face, and how to find and work with a lawyer. The workbook ($15 for individuals; $20 for public interest organizations; $40 for libraries, institutions, law firms, and government agencies) is available from **Environmental Action Foundation** (1526 New Hampshire Ave. NW, Washington, DC 20036; 202-745-4870).

Another helpful guide is *A Citizen's Toxic Waste Audit Manual* by Ben Gordon and Peter Montague. The guide is designed to help you identify toxic pollutants generated by your community; find out who is importing toxic wastes into your community; pressure local polluters to reduce waste; and fight polluting facilities proposed for your community. The manual describes the process of conducting a waste audit—"an assessment of the types and amounts of chemical wastes that are being emitted by a particular facility to a particular community." The guide is available from **Greenpeace U.S.A. Toxics Campaign** (1017 W. Jackson Blvd., Chicago, IL 60607; 312-666-3305). The 72-page guide is free, although Greenpeace requests a $5 donation to defray expenses.

Other Ways to Get Involved

Lawsuits against local polluters are only one type of action that environmentally concerned citizens can take. Here are ten additional ways to get involved in environmental issues.

- **Monitor legislation.** All of the key environmental organizations keep tabs on state and national laws being considered in order to offer testimony and generate letter-writing campaigns on behalf of (or against) proposed laws. Three national organizations have established telephone legislative hotlines to keep interested individuals apprised of bills before Congress: the **Audubon Society** (202-547-9017), **Sierra Club** (202-547-5550), and the **National Wildlife Federation** (202-797-6655). Each hotline offers a two- to four-minute recording.

• **Write letters.** It may not seem like a potent weapon, but letters to state and federal legislators on pending bills *do* influence their opinions. According to the National Wildlife Federation, the largest conservation organization in the United States, letter writing helped pass both the 1988 Endangered Species Act and the Clean Water Act. By keeping in touch with local and national environmental organizations, you will know which legislators to write to about what subject, thereby making sure your letters are timely and have maximum impact. Write to senators at the U.S. Senate, Washington, DC 20510; write to members of the House at the U.S. House of Representatives, Washington, DC 20515. Write to the president at the White House, Washington, DC 20500. Use a public library for names and addresses of state and local officials.

When writing to any public official, keep your letter simple. Focus on one subject and identify a particular piece of legislation, if appropriate. Request a specific action ("Please vote in favor of SB 26890") and state your reasons for your position. If you live or work in the legislator's district, make sure to say so. If you write to a legislator from another district, send a copy of the letter to your own legislator. Keep the letter to one or two paragraphs, never more than one page. Unemotional, courteous letters work best. You might send a copy of the letter to a local newspaper.

In addition to writing letters, you might telephone the office of your legislator. All members of Congress have one or more local offices in their home district, although it probably is more effective to telephone their main office in Washington, DC (The main switchboard for Congress is 202-224-3121; operators can connect you to any House or Senate office.) Don't expect to talk to the legislator directly, of course. Ask to speak with the person monitoring environmental issues, state your position in under one minute, thank the individual for speaking with you, and say good-bye.

One handy resource is the *Activist Kit* from **National Wildlife Federation** (1400 16th St. NW, Washington, DC 20036; 202-797-6800). The $5.95 booklet provides information helpful in contacting members of Congress and officials in federal agencies on conservation issues.

• **Educate others.** You can do this in a variety of ways, from talking to your friends, co-workers, and neighbors to organizing an educational activity. One such activity might be a "stream walk," a group hike during which participants try to diagnose potential problems. Walking a stream with a knowledgable leader can alert you to erosion problems, highway and construction runoff, excessive algal growth, poisoned fish, foul smells, and direct discharges into the stream. Each of these is cause for concern. The stream walk might be followed by a stream cleanup. Be sure to invite reporters from newspapers and radio and television stations to these activities; their reports will help educate others.

• **Campaign for environmental candidates.** Don't just be concerned about someone claiming to be an "environmental president." Look at the environmental positions of candidates at all levels of government. In fact, local candidates—mayor, city council, county supervisors—will probably have a more immediate impact on policies and programs that directly affect the environment in your area. On the national level, the **League of Conservation Voters** (LCV, 1707 L St. NW, Ste. 550, Washington, DC 20036; 202-785-VOTE) publishes the National Environmental Scorecard, rating state and national candidates and office holders for their environmental records. LCV also has several state chapters that track local candidates.

• **Launch a campaign at work or school.** At Rutgers University, for example, members of the law association decided to target the use of plastic foam in the cafeterias. After creating a multistep, long-term strategy, the students first approached the food services department. The director of food services readily committed to getting rid of foam cups in a matter of days, and the foam food containers as soon as the current inventory was depleted. One of the lessons the students learned: Sometimes, all you have to do is ask. Another school-based activity took place at Brown University, where students saved the school about $2,000 by promoting energy conservation and awareness.

The workplace—whether an office or factory—also offers many opportunities for promoting environmental issues. Recycling paper, bottles, cans, cardboard, and other materials is one good place to start. Removing undesirable materials from the company cafeteria or snack bar is another good effort. Getting the entire company to take on a pet environmental project—cleaning up a local stream, for example, or creating a global-warming campaign—can educate both employees and the community at large. (See also "Making Your Business Green.")

• **Promote community recycling.** One of the most common ways to form a recycling organization is to organize a meeting of friends, neighbors, concerned individuals, and members of civic groups interested in recycling. You can hand out fliers throughout the neighborhood and post notices in public spaces to invite people to attend your meeting.

To promote existing recycling opportunities in your community, you can organize a Recycling Day to bring people to the local recycling center. You can also make presentations to local schools and civic groups to promote existing recycling opportunities in your community. Advertisements or articles in local newspapers can facilitate more involvement in local recycling activities. (If you are not sure of the location of the nearest recycling center, check the Yellow Pages under "Recycling Centers" or contact a local trash-hauling company—listed in the yellow pages under "Rubbish and Garbage Removal.")

To start a neighborhood recycling center, you need to speak with local salvage and recycling companies, representatives of the beverage industry, and local garbage haulers. All of these groups can help facilitate the collection and transportation of recyclables. Through lobbying, public education, and media work, community recycling groups can be effective advocates of city- or county-operated curbside recycling programs. For more information on starting a community recycling organization, send for the Greenpeace Action Community Recycling Start-up Kit, available from **Greenpeace Action**, 1436 U St. NW, Washington, DC 20009; 202-462-1177. Another helpful resource is "Why Waste a Second Chance?" a motivational video and accompanying guide-

Stamp Out Greenwashing!

Maybe you're mad as hell about Company X making dubious environmental claims about its product. Maybe you're the president of Company Y, which just happens to be Company X's main competitor, and think that Company X's ads are full of hot air. Either way, instead of just fuming at some magazine page or TV screen, you can get help from the National Advertising Division (NAD) of the Council of Better Business Bureaus.

NAD's mission is to investigate ad claims—both on its own and as a result of complaints by other companies and just plain folks. Increasingly, many of those complaints have to do with green marketing. In the last year, of the 88 claims NAD investigated, 16 were about environmental claims. In about a third of the cases, the division sided with the company that originally made the claim; in two-thirds of the cases, the offending companies changed or dropped their claims.

"It's not just limited to advertising. We also do labels," says Ron Smithies, NAD's director. He says the division has addressed claims that products were environmentally friendly, recycled, recyclable, or biodegradable. "After you've done 16 cases you start to recognize patterns."

There may be a ray of hope in all this from manufacturers and their mouthpieces on Madison Avenue: Smithies says the ad industry approached NAD last summer, asking it to help the Federal Trade Commission develop guidelines about the use of environmental terms in packaging and advertising. NAD's function is to generate some real-life cases to support guidelines the FTC might someday propose. So, there may well be a day ahead when words like nontoxic, natural, and environmentally safe actually mean something.

If you have a complaint about an ad, "Just tear it out and send it to us," says Smithies. "That's the best way to get us interested." TV and radio ads can be challenged, too. Include the time and date of broadcast, or the issue and page of publication.

Send the information to the National Advertising Division, Council of Better Business Bureaus, 845 Third Ave., New York, NY 10022; 212-754-1320.

book produced by the **National Association of Towns and Townships** (1522 K St. NW, Suite 730, Washington, DC 20005; 202-737-5200). The video is intended to help community leaders and concerned citizens generate interest and support for recycling.

• **Plant trees.** We've already discussed the many benefits of trees on your community and on the environment. (See "How Trees Save the Earth.") Setting up a tree-planting program is relatively easy. There are a number of organizations that can help organize a one-day or longer-term community campaign to educate citizens on a wide range of environmental issues while assisting them in finding suitable sites, then planting and caring for saplings. Contributors of $15 or more to **Global ReLeaf** (c/o American Forestry Association, 1516 P St. NW, Washington, DC 20005; 202-667-3300) receive a free *Global ReLeaf Action Guide*, containing background information, an action checklist, and access to other Global ReLeaf resources.

• **Invite speakers to your organization.** Most environmental organizations offer speakers on a wide range of topics who will speak at no charge to your civic, school, religious, or social organization. Many will also provide literature and additional information. For maximum impact, consider scheduling a debate or panel discussion among representatives of environmental groups, government agencies, and industry. You might turn a seemingly simple topic into a lively discussion. Be sure to invite local media to any such event to increase the exposure of your message.

• **Get involved with government.** Most communities offer a variety of boards, commissions, and committees that deal with environmental issues: planning commissions, zoning and land-use commissions, parks commissions, transit boards, and so on. Each can play a role in setting policies that affect the quality of the environment in your area. For information on how to participate in any of these organizations, contact such organizations directly. Most such groups encourage participation by as many citizens as possible. For information on specific agencies and organizations in your area, contact the local branch of the League of Women Voters or any local or national environmental organization.

THE POWER OF CONSUMER BOYCOTTS

Sometimes, creating change requires more than individual actions. Sometimes, it takes some joint efforts.

Organized boycotts against companies have been known to work wonders. During the past few years, boycotts have become one of the tools of choice of groups whose interests span the political

Class Actions

Today's students don't seem to need a lot of incentive to get involved with environmental issues—they seem to gravitate to it naturally—but they often do need inspiration and direction. Two books should fit the bill quite nicely:

The Student Environmental Action Guide: 25 Simple Things We Can Do was written by the Student Environmental Action Coalition (SEAC), a national membership organization. Their handy 96-page guide, written in the same friendly style as Earth Works' *50 Simple Things You Can Do* series, brings the environment onto the college campus. (Although there is a nod to high schoolers, a large portion of the projects relate to a college setting.)

Many of the other steps are predictable—recycling, eliminating disposable cups, planting trees, carpooling, switching to compact fluorescent light bulbs, and so on—but that does not make them any less valuable. By making these actions so accessible to such an important audience, this book is destined to become as much a part of the college culture as midterms and finals.

Copies ($4.95) are available in bookstores, or from SEAC, P.O. Box 1168, Chapel Hill, NC 27514.

Campus Ecology: A Guide to Assessing Environmental Quality & Creating Strategies for Change, by April A. Smith and SEAC calls itself "a guide for analyzing the nature and magnitude of environmental issues on campus." The first two sections of the book, "Wastes and Hazards" and "Resources and Infrastructure" outline how to research and analyze campus environmental practices, issue by issue. The third section, the "Business of Education," addresses the environmental impacts of a school's research and economic activities and its ethical stance. The final section, "Taking Action," provides strategies for creating and implementing environmentally responsible practices.

The attractive 144-page paperback sells for $17.95 in bookstores, and is available from SEAC at the above address.

spectrum: prolabor, antilabor, prochoice, right to life, antinuclear energy, antigun control, etc. And, of course, the environment.

"It seems that the Exxon *Valdez* oil spill reminded the American public that it has the option to boycott," says Todd Putnam, editor of *National Boycott News*, a sporadically published report that tracks boycotts. While not all boycotts work, and some are indeed obscure, a significant number hit their mark, forcing companies to reconsider their products, packages, or policies.

Effective boycotts take time, a factor that frustrates some individuals who seek instant action. Perhaps the most successful environmentally related boycott of recent years was against tuna canners, who had fished for tuna in a way that also killed dolphins. In April 1990, when the three major tuna canners announced that they would change their dolphin-killing ways, it was the culmination of years of efforts, according to Brenda Killian, associate director of the Earth Island Institute's Save the Dolphin Project, which spearheaded the boycott.

The tuna boycott worked despite the fact that the number of American consumers actively boycotting tuna was believed to be relatively small—a million or so people, according to most estimates; no one knows exact numbers. That's well under 1% of the marketplace. In fact, the tuna industry was scarcely affected by the actions: during the 12 months preceding the companies' announcements, tuna sales were relatively unchanged.

Sprout Clout

That companies are listening—and responding—to boycotters isn't surprising when you consider who's involved. A 1989 Roper survey found that the demographics of boycotters represent many companies' prime markets: college-educated, thirtysomething, two-income families. That makes boycotters a potent group.

Perhaps even more potent are environmentally conscious kids—"green sprouts," as they are known. "Kids are suddenly looking around and not just taking everything for granted, but realizing that clean water and trees aren't always going to be here," says Karen Firehock of the Izaak Walton League. "It leads them into becoming questioners instead of accepters."

Brenda Killian agrees that children are an effective weapon in boycotts. In fact, she says, "Many CEOs have had to face their own children asking, 'Why are you killing dolphins?' Children respond quicker to a call for boycotts. They write what they really feel."

No one knows this better than McDonald's Corporation, which found itself confronted with angry letters from kids across the country about its use of foam hamburger boxes and other nonrecyclable waste. The letters, organized in part by the New Jersey–based Kids Against Pollution, persuaded the company to renounce polystyrene in favor of paper, despite the company's insistence that foam is better for the environment. Again, while there is little reliable data on the number of young boycotters, it is thought to be relatively small. But, according to company officials, the idea of a generation of Americans growing up with a bad taste in their mouths about Big Macs and McNuggets was too persuasive for them to ignore.

From Timber to Tourism

What else is being boycotted? The list of causes and companies is long. *National Boycott News*, which tries to publish four times a year, contains nearly 200 dense pages of details on boycotts. (Subscriptions are $10 a year for individuals, $20 for nonprofit groups, and $40 for others.) Among the issues to be included will be boycotts against entire industries. Producers of lamb and wool are on the list, for example, because of sheep-ranchers' use of a compound that threatens wildlife. Makers of stonewashed and acid-washed clothing are targets, too, because the pumice they use to make clothes look worn is strip-mined in the Jemez Mountains in New Mexico. Alaskan tourism is on the hit list because of the state's support of wolf hunts in certain parts of the state.

The list goes on. Chrysler, Cadillac, Honda, and Oldsmobile are being targeted because of their alleged refusal to use recycled oil in their production lines, or to recommend it to car buyers. Mitsubishi International Corporation has been targeted because one of its companies is engaged in tropical deforestation in Malaysia. Rainforest issues are high on boycotters' lists. According to Victor Menotti of the Rainforest Action Network, two paper and lumber companies, Weyerhaeuser and Georgia-Pacific, are also on the list; both have sold off their tropical timber operations, but are still involved in distribution of tropical timber products.

In the end, no company is sacred. But the point isn't to boycott everything that isn't perfect. As in life itself, it's most effective to choose your fights, voting with your dollars—and letters—on issues about which you care deeply.

Another good source of information on boycotts is **Co-op America**, which publises "Boycott Action News" in each issue of its *Co-op America Quarterly*. For a free sample copy and membership information contact Co-op America, (2100 M St. NW, Ste. 3190, Washington, DC 20036; 202-872-5307).

INVESTING IN THE ENVIRONMENT

If you care about the earth, are careful about the energy you use, and think about the products you buy, it would follow that you might want your investments to reflect these same ideals. Indeed, putting your money into socially responsible or environmentally conscious investment vehicles may bring a decent rate of return, but a lot of your investment decisions depend on your specific objective: Do you want to make money, help clean up the environment, or simply park your dollars in a place that has minimal environmental impact?

Before we proceed with some answers, let's dispense with the basic disclaimers. We're not investment advisers and nothing in this article (or any other article we write) should be taken as investment advice. You should always consult with your own professional investment counselors and read the literature issued by prospective investment companies before handing over your money.

There is a growing list of mutual funds now on the market offering opportunities for environmental investing, but beware: Not all eco-funds are created equal. As with so many other things environmental, there are many shades of "green." In this case, there are two major types of environmental funds: mutual funds and sector mutual funds. The first type of funds hold stock in companies that have good environmental records in their operations or products, even though their business need not be tied directly to the environment.

Sector funds, the second type, buy stock in firms directly involved in environmental cleanup technology, pollution control devices, wastewater treatment, and waste hauling. This may sound laudable—after all, why not help those companies that are trying to help the environment?—but things aren't that simple. Some waste handling companies have poor environmental records. For example, Waste Management, Inc. (WMI), and Browning-Ferris Industries (BFI), the nation's two largest waste-handling firms, have extensive records of criminal and civil wrongdoing. For starters, in 1990, WMI paid $3.75 million to settle EPA charges that the firm had overstuffed its Chicago-area hazardous waste incinerator, disconnected monitoring equipment, and routinely

Socially Responsible Funds

Here are names of leading environmental and socially responsible mutual funds, along with their year of founding and telephone number. You are encouraged to contact these funds to obtain their latest prospectus.

Calvert Ariel Appreciation	1990	800-368-2748
Calvert Ariel Growth	1986	800-368-2748
Calvert Social Investment Bond	1987	800-368-2748
Calvert Social Investment Equity	1987	800-368-2748
Calvert Social Investment Growth	1982	800-368-2748
Covenant Portfolio	1992	800-833-4909
Domini Social Index Trust	1991	800-762-6814
Dreyfus Third Century Fund	1987	800-782-6620
Green Century Balanced Fund	1992	800-934-7336
Merrill Lynch Eco-Logical Trust	1990	800-422-9001
Muir California Tax Free Fund	1991	800-648-3348
New Alternatives	1982	516-466-0808
Parnassus Fund	1985	800-999-3505
Pax World	1971	800-767-1729
Righttime Social Awareness	1990	800-242-1421
Schield Progressive Environmental Portfolio	1990	800-826-8154
Working Assets Money Market	1983	800-533-3863

spilled PCBs without reporting the spills. In 1991, an explosion at the incinerator forced its closure. WMI also agreed to pay $19.5 million to settle a lawsuit over price-fixing; BFI was also involved in the case. While agreeing to pay fines, both companies denied wrongdoing. So cleanup funds aren't always all that clean.

A third type of fund is a hybrid: an environmental sector fund that applies social responsibility screens to its companies. One such fund is the Schield Progressive Environmental Fund.

Most socially responsible investors don't want to buy into funds that own shares in companies that pollute. These are for people, in the words of Mindy Lubber, manager of the Green Century Fund, who want to invest and "do it in a way that's more consistent with who I am personally." Her fund was started by several nonprofit public-interest groups that discovered that mutual funds in which they had invested had holdings in tobacco companies and producers of toxic chemicals, among other less-desirable firms. "They wanted to be more consistent," Lubber says. "Some of the companies that were the greatest threats to the environment were in their mutual fund portfolios."

Doing Due Diligence

The first question in any investor's mind ought to be whether a fund brings a good rate of return. This may not be important to everyone, but at least a fund shouldn't lose money. There's money to be made in environmental investing, but you must pay attention to several key factors if you expect to actually see a profit.

Investing your money in any one particular industry can increase your risk, so choosing funds that invest in companies in several industries may be a consideration. (For that matter, putting all your money into stocks—as opposed to bonds, cash reserves, certificates of deposit, collectibles, or other investments—can be risky, too. That's where investment counseling is key.)

In judging environmental funds, you must apply the same scrutiny and criteria as you would when investing in any other fund. Marilyn Ericksen, head of the socially responsive investing group of the firm Scudder Stevens & Clark, says, "You can develop a very solid portfolio based on socially responsible companies. You start with the company. Is the company well managed? Is it financially sound? Then you apply the social screens. If the company doesn't pass them, you don't buy it. You look for another one that meets your criteria. You can always find companies that are good investments and that meet your social screens."

There's another important consideration: Most environmental mutual funds are so new that they don't have long-term track records. This is key in determining a fund's performance during both boom times and bust. That lack of a history increases the risk level of some funds, although it by no means rules them out.

So you must read the literature and figure out what's a good investment for your needs, weighing the type of fund against its performance. There are different places to put your money, and a seemingly endless array of investment theories, including those that go well beyond the social and environmental implications of investments. For example, you must also consider the aggressiveness of the funds—the way each trades off risk for a potentially higher rate of return. The Pax World Fund and the Calvert Social Investment Managed Growth funds are known as "balanced funds," because they hold both stocks and bonds. They will allocate investments, based on their appraisal of the economy, among cash, common stock, and bonds. For example, the Green Century fund holds 60 percent of its assets in stock, 35 percent in bonds, and 5 percent in cash. Other funds, such as the Dreyfus Third Century Fund, hold only stocks, and so are called "equity funds."

In the end, you must balance environmental ideals with sound business sense. After all, we're talking about your hard-earned dollars here, perhaps even your retirement. There's no getting around reading the literature, following funds' performance, and getting advice from professionals.

Sources of Investment Information

• *The Better World Investment Guide*, by Myra Alperson, Alice Tepper Marlin, Jonathan Schorsch, and Rosalyn Will, $22.95 ppd. from the Council on Economic Priorities, 30 Irving Pl., New York, NY 10003; 800-822-6435. Also available: *Shopping for a Better World*, which rates hundreds of companies on 11 socially responsible criteria. ($7.49 per copy, including shipping.)

• *A Socially Responsible Financial Planning Guide*, $2.50 ppd. from Co-op America, 2100 M St. NW, Ste. 310, Washington, DC 20036; 202-872-5307.

• The Social Investment Forum (430 First Ave. N., Minneapolis, MN 55401; 612-333-833, a clearinghouse of information on social investing, publishes *The Forum* newsletter ($35/year with membership) and the *Social Investment Services Guide* ($45; $30 for members).

MAKING YOUR BUSINESS GREEN

The idea that what's good for the environment is good for business is not exactly new, but it is one that has gained considerable adherents in recent years. Previously limited to smaller companies like **Ben & Jerry's, Tom's of Maine,** and **The Body Shop,** the list of environmentally responsible firms includes such Fortune 500 companies as **Clorox, S.C. Johnson Co.,** and **H.P. Fuller Co.**

How to Be a Greener Company

Like people, each company is unique. And just as each individual has different ways of "going green," companies, too, have a variety of options available to them. What's right for one company may not work for another.

Just as we tell people not to try to be perfectly green — lest they get frustrated and discouraged at the enormity of it all — we tell the same things to companies. It's hard to do everything right, even if you are making an environmentally benign product or performing a service that doesn't seem to pollute a thing. Just being in business is a polluting activity. There is so much paper, so many purchases, so much energy and other resources needed just to operate from day to day.

Here are a few things companies can do to be more environmentally responsible. For further reference, consult *50 Simple Things Your Business Can Do to Save the Earth*, by the Earthworks Group, available for $8.45 postpaid from Tilden Press, 1526 Connecticut Ave. NW, Washington, DC 20036.

• **Conduct a waste audit.** A lot of the things companies throw away are recyclable. Paper, of course, which makes up about 41 percent of landfills, is the first thing to look at. Office paper use has grown enormously in recent years, despite repeated promises of a paperless "office of the future." But it has become painfully obvious, as one expert put it, that the paperless office is about as practical as a paperless bathroom. But paper is only the beginning. Cardboard boxes, packaging materials, aluminum, glass, steel, and some plastics all can be saved from the waste stream.

There is good incentive for companies to do these things. Many companies are earning a tidy sum selling the "trash" they used to throw away. (In fact, these companies once paid to have it hauled away by trash haulers; now they get paid for it from recyclers.)

• **Consider the kitchen.** Just like your kitchen at home, your office kitchen or cafeteria offers many opportunities for cutting waste. Those styrofoam coffee cups are a great place to start. Even a small office or department can go through mountains of them every week, as everyone grabs a new one several times a day. By replacing them with ceramic (or even plastic) mugs, you'll save trash and cut costs. Everything else, from the type of tuna purchased for the company cafeteria (is it dolphin-safe?) to the amount of disposable utensils used, provides an opportunity for reducing unnecessary trash.

• **Get people out of their cars.** A green company will encourage employees to carpool and use public transportation. There are many ways to do this. Some companies give preferred parking spaces to carpoolers. Others offer financial incentives to take the bus, train, or subway. Still others offer free shuttle vehicles to bring employees to and from public transportation.

• **Cut energy use**. There are dozens of ways to do this in offices and factories, from new lighting fixtures and bulbs that provide more light for less electricity, to window glazings that can significantly cut heating and air-conditioning use. The AC system itself is a big energy drainer, and keeping it maintained can save both energy and money.

• **Educate everyone.** Companies are looked to by employees, customers, and neighbors in the community for leadership. Many companies are using their advertisements, annual reports, brochures, and other communications to help educate others on one or more environmental topics. Other companies are sponsoring events, or creating teaching units on the environment for classroom use. By becoming a trusted information source on environmental issues, companies can gain credibility and commitment from those both inside and outside the organization.

As you begin to see, many of these environmentally responsible moves also can pay off at the bottom line, whether directly through increased sales or indirectly through employee motivation and customer loyalty. And that makes green businesses even greener.

The Ceres Principles

In 1989, a coalition of environmental- and social-responsibility groups organized the **Coalition for Environmentally Responsible Economies** (or Ceres, after the Roman goddess of agriculture). Ceres' first act was to publish the Ceres Principles (formerly called the Valdez Principles), a set of standards "for evaluating activities by corporations that directly or indirectly impact the earth's biosphere." The idea of the principles is "to create a voluntary mechanism of corporate self-governance that will maintain business practices consistent with the goals of sustaining our fragile environment for future generations, within a culture that respects all life and honors its interdependence."

Here is a summary of the Ceres Principles:

- **Protection of the biosphere:** Companies will minimize the release of any pollutant that may damage the air, water, or earth, including those that contribute to the greenhouse effect, depletion of the ozone layer, acid rain, and smog.
- **Sustainable use of natural resources:** Companies will make sustainable use of renewable natural resources, such as water, soils, and forests, including protection of wildlife habitat, open spaces, and wilderness, and preservation of biodiversity.
- **Reduction and disposal of waste:** Companies will minimize waste, especially hazardous waste, and recycle whenever possible. All waste will be disposed of safely and responsibly.
- **Wise use of energy:** Companies will make every effort to use environmentally safe and sustainable energy sources and invest in energy efficiency and conservation.
- **Risk reduction:** Companies will minimize environmental and health and safety risks to employees and local communities by employing safe technologies and by being prepared for emergencies.
- **Marketing of safe products and services:** Companies will sell products or services that minimize adverse environmental impact and that are safe for consumer use.
- **Damage compensation:** Companies will take responsibility for any harm caused to the environment through cleanup and compensation.
- **Disclosure:** Companies will disclose to employees and com-

All This and a Paycheck, Too

Green At Work: Finding a Business Career that Works for the Environment, by Susan Cohn, is a valuable book for anyone interested in doing well by doing good, offering a wealth of insight and inspiration on how to find an environmentally related job in both the private and nonprofit sectors.

At the heart of the book is a section on career profiles and another called "The Corporate Directory." The profiles consist of 30 or so first-person job descriptions from individuals working in environmental positions, mostly in the private sector. Among those featured are some of the leading authorities of green marketing, management, and consulting, at such firms as American Express, Arthur D. Little, Church & Dwight, Colgate-Palmolive, Coopers & Lybrand, S. C. Johnson & Son, and Merrill Lynch. Each one- to two-page profile, distilled from interviews conducted by Cohn, a career counseler at New York University's Stern School of Business, offers an inside peak not just at what each individual does, but how he or she got there.

The Corporate Directory is where most readers will turn first. It offers names, addresses, and contacts at more than 200 companies that have environmental programs, policies, products, or services. Having such information will no doubt short-circuit the time-consuming "networking" process for thousands of individuals.

There's more: lists of organizations, government agencies, publications. All told, Cohn packs a lot into 232 pages. Short of holding your hand and taking you to a job interview, it's hard to imagine a more useful resource for job-seekers.

Green At Work is $16 in paperback from Island Press (1718 Connecticut Ave. NW, Ste. 300, Washington, DC 20009; 202-232-7933).

munity incidents that cause environmental harm or pose health or safety hazards.

• **Environmental directors and managers:** At least one member of the board of directors will be qualified to represent environmental interests, including funding a senior executive position for environmental affairs.

• **Assessment and annual audit:** Companies will conduct annual self-evaluations of progress in implementing these principles and make results of independent audits public.

For more information about the Ceres Principles, contact the **Social Investment Forum** (711 Atlantic Ave., Boston, MA 02111; 617-451-3252).

ENVIRONMENTAL ORGANIZATIONS

Acid Rain Foundation (1410 Varsity Dr., Raleigh, NC 27606; 919-828-9443) is dedicated to developing public awareness, information, educational materials, and research in the area of acid disposition, air pollution and toxins, global change, and recycling. Membership is $25 per year and includes the quarterly publication, *The Acid Rain Update*.

Acid Rain Information Clearinghouse (46 Prince St., Rochester, NY 14607; 716-271-3550) provides comprehensive reference and referral, current awareness, and educational services to a wide range of professionals, academics, and public interest groups. The organization maintains a library, sponsors conferences and seminars, and prepares topical bibliographies.

Adopt-a-Stream Foundation (P.O. Box 5558, Everett, WA 98206; 206-388-3313) promotes environmental education and stream enhancement. It offers support and guidance to those "adopting" a stream: The adopting group provides long-term care of the stream. Membership in this nonprofit, volunteer-run organization is $10 and up, and includes a subscription to the quarterly newsletter, *StreamLines*. The foundation publishes two informative books: *Adopting a Stream* and *Adopting a Wetland*.

African Wildlife Foundation (1717 Massachusetts Ave. NW, Washington, DC 20036; 202-265-8394) works with Africans in over twenty countries, promoting, establishing, and supporting grassroots and institutional programs in conservation, wildlife management training, and management of threatened conservation areas. The group's current emphasis is on educating Americans not to buy ivory. Membership is by donation; $15 minimum required to subscribe to the quarterly newsletter, *Wildlife News*.

Alliance to Save Energy (1725 K St. NW, Ste. 509, Washington, DC 20006; 202-857-0666) is dedicated to increasing energy efficiency. The group conducts research and pilot projects to evaluate solutions to energy-efficiency problems. Alliance programs address quality of life, environment, national security, international competitiveness, and economic development. It is a nonmembership organization supported by corporations, foundations, and other organizations promoting energy efficiency. Publications and computer software are available to the public.

America the Beautiful Fund (219 Shoreham Bldg., Washington, DC 20005; 202-638-1649) was organized in 1965 to give recognition, technical support, small seed grants, gifts of free seeds, and national recognition awards to volunteers and community groups that initiate new local projects improving environmental quality. Activities include beautifying communities through seed donations and growing food for the needy. Membership ($10 and up) includes a subscription to *Better Times*, a quarterly newsletter. Seeds cost $4.95, which includes shipping and handling.

American Cave Conservation Association (P.O. Box 409, Horse Cave, KY 42749; 502-786-1466) was established to protect and preserve caves, karstlands, and groundwater. The group work focuses on education, creation of a national education facility and museum of caves and karst, and developing an information network. Annual membership fees are $25 and include a bimonthly magazine *American Caves* and a periodic newsletter.

American Cetacean Society (P.O. Box 2639, San Pedro, CA 90731; 310-548-6279) is a nonprofit volunteer organization working to protect whales and dolphins through research, conservation, and education. Projects have focused on the killings of dolphins by tuna fishermen, stopping whaling, ocean pollution, strandings, and gill net and drift net problems. Membership is $25 per year and includes a subscription to the quarterly *Whale Watcher* magazine and *Whale News* newsletter.

American Council for an Energy-Efficient Economy (1001 Connecticut Ave. NW, Ste. 801, Washington, DC 20036; 202-429-8873) gathers, evaluates, and disseminates information to stimulate greater energy efficiency. Topics of focus include buildings, appliances, and indoor air quality. It annually publishes the booklet *The Most Energy Efficient Appliances*. Although not a membership organization, it maintains a mailing list.

American Forest Council (1250 Connecticut Ave. NW, Ste. 320, Washington, DC 20036; 202-463-2455) is sponsored by the forest products industry to educate and train private forest land owners in good forest-management practices. It sponsors an education program, "Project Learning Tree," through its American Tree Farm System. The council publishes the monthly *American Forest Council* magazine; a poster/magazine *Green America* is available through Project Learning Tree. Membership is available to companies, private forest tree farmers and environmental and conservation organizations.

American Forest Association (P.O. Box 2000, Washington, DC 20013; 202-667-3300) is a national citizens organization dedicated to the maintenance and improvement of the health and value of trees and forests, and to make Americans more aware of and active in forest conservation and tree planting. It sponsors "Global ReLeaf," a national program encouraging Americans to plant millions of trees to lower carbon dioxide levels and beautify communities. Another major program is the "National Register of

Big Trees." Publications include the *Resource Hotline* newsletter. Membership is $24 annually and includes a subscription to the bimonthly *American Forests*.

American Humane Association (63 Inverness Dr., East, Englewood, CO 80112; 303-792-9900) was founded to protect animals from neglect, abuse, and exploitation. In the late nineteenth century, after the first case of child abuse was successfully prosecuted, it added child welfare work to its mission and established the Children's Division of the organization. Animal Protection Division activities include training programs for animal care and control professionals. The group also provides humane education materials and methods for children and adults, emergency relief for animal victims of natural disasters, advocacy for humane legislation and safeguards for animal actors, and public policy to protect abused and neglected families. Membership starts at $15 and includes the quarterly magazine *Advocate*.

American Littoral Society (Sandy Hook, Highlands, NJ 07732; 908-291-0055) is a nonprofit conservation organization founded to study and conserve the littoral zone, the fragile, productive areas where the sea meets the shore. Many local chapters sponsor international field trips, beach cleanups, tern nest patrols, and turtle nest watches. Annual dues are $25 and include the quarterly newsletter *Underwater Naturalist* and the bimonthly newsletter *Coastal Reporter*.

American Rivers (801 Pennsylvania Ave. SE, Ste. 400, Washington, DC 20003; 202-547-6900) is a nonprofit group working to preserve the nation's rivers and their landscapes. American Rivers measures its progress in river miles preserved, streamside acres protected, dams blocked, and taxpayer dollars saved. Membership dues are $20 and up annually. Members receive the quarterly newsletter *American Rivers*, along with a list of outfitters that support river conservation.

American Society for Environmental History (Center for Technology Studies, New Jersey Institute of Technology, Newark, NJ 07012; 201-596-3291) promotes research, publications, teaching, and communications on the relationship of humans to the natural environment from a broadly historical and humanistic perspective. Membership ($24 individuals, $12 students, $30 institutions) includes the quarterly journal *Environmental Review*.

American Society for the Prevention of Cruelty to Animals (441 E. 92nd St., New York, NY 10128-6699; 212-876-7700) was the first humane society in America. Its purpose is to provide effective means for the prevention of cruelty to animals throughout the United States. The group focuses on the rights of companion animals, animals in research and testing, animals raised for food, wild animals, entertainment and work animals, and animals in education. It also sponsors a yearly safari to Kenya to observe wildlife in a natural setting. Membership ($20 per year) includes the *ASPCA Quarterly Report* magazine and a discount on pet products in its New York store.

American Wildlands (40 East Main St., Ste. 2, Bozeman, MT 59715; 406-586-8175) is a national conservation organization which promotes ecologically sustainable uses of public wildland resources, including forests, wilderness, wildlife, fisheries, and rivers, principally in the Interior West. Offices in Bozeman, MT; Denver, CO; and Reno, NV. Membership ($25 and up) includes the quarterly newsletter *On the Wild Side.*

Americans for Safe Food (see **Center for Science in the Public Interest**)

Americans for the Environment (1400 16th St. NW, Box 24, Washington, DC 20036; 202-797-6665) is an educational organization established to train, educate, and involve the environmental community in issues and strategy, and to encourage activists to become involved in campaigns to elect environmentalists to public office. This is not a membership organization.

Animal Protection Institute of America (2831 Fruitridge Rd., Sacramento, CA 95820; 916-731-5521) was established to eliminate or alleviate fear, pain, and suffering among animals through humane education and member action. Membership ($20 per year) includes a subscription to the quarterly magazine *Mainstream.*

Animal Welfare Institute (P.O. Box 3650, Washington, DC 20007; 202-337-2332) promotes the welfare of all animals and works to reduce the pain and fear inflicted on animals by humans. One focus is on improving conditions of laboratory animals, "factory farmed" animals, and species threatened by extinction. It maintains an extensive collection of publications and films for educational use. Membership ($15 annually) includes *Animal Welfare Institute*, a quarterly newsletter, along with options on free books when they become available and invitations to events at the Institute.

Atlantic Center for the Environment (39 S. Main St., Ipswich, MA 01938; 508-356-0038) was established to promote environmental understanding and encourage public involvement in resolving natural resource issues in the Atlantic Ocean region. The center has worked on studying migratory birds, river, and watershed management, and various international programs. Annual dues ($25 and up) include a subscription to the organization's quarterly magazine *Nexus.*

Basic Foundation (P.O. Box 47012, St. Petersburg, FL 33743; 813-526-9562) was established to promote efforts to balance population growth with natural resources and tropical rain forest preservation. The foundation supports research activities, exhibits, publications, conferences, lectures, and nature tours. The foundation also publishes and donates educational materials to schools; lobbies policy-makers and international organizations on behalf of the tropical rain forests; sells products with environmental messages; supports activities of various rain forest projects; and sponsors nature tours to rain forests in Costa Rica.

Bat Conservation International (P.O. Box 162603, Austin, TX 78716-2603; 512-327-9721) educates about the vital role of bats in world environments. Bats are responsible for controlling insects, pollinating plants, and generating up to 95 percent of the seed dispersal essential to the regrowth of cleared tropical forests. Membership is $30 and up annually and includes the quarterly newsletter *Bats*. *America's Neighborhood Bats* is offered through the organization's catalog.

Bio-Integral Resource Center (P.O. Box 7414, Berkeley, CA 94707; 510-524-2567) is dedicated to providing information on least-toxic pest control. Membership fees depend on subscriptions desired: the *IPM Practitioner*, published ten times per year, is $25; the *Common Sense Pest Control Quarterly* is $30; both are available for $45. Members can receive help with pest-management problems.

Caribbean Conservation Corp. (P.O. Box 2866, Gainesville, FL 32602; 904-373-6441) focuses on marine and sea turtle research and conservation. It operates the Green Turtle Research Station in Tortuguero, Costa Rica, and maintains a green turtle tagging project in cooperation with the Center for Sea Turtle Research at the University of Florida. It also runs the Volunteer Research Travel Program, with spring and summer trips to Tortuguero; participants assist research teams in studying and tagging marine turtles. Membership is $35 a year and includes the quarterly newsletter *The Velador*.

Center for Environmental Information (46 Prince St., Rochester, NY 14607; 716-271-3550) is an information organization that maintains an extensive library, organizes conferences and seminars, and publishes many useful books, manuals, and directories. The center also provides a discount travel club for members. Membership ($25 and up per year) includes a bimonthly update of the center's activities (*Sphere*), invitations to monthly "timely topic" seminars, and other publications.

Center for Investigative Reporting (530 Howard St., 2nd Fl., San Francisco, CA 94105-3007; 415-543-1200) provides support for investigative journalism, including environmental stories, for international television networks, newspapers, and magazines. The center has a regional office in

Washington, DC. Environmental projects include a documentary for public television focusing on using the earth as a dumping ground for toxic wastes.

Center for Marine Conservation (1725 DeSales St. NW, Ste. 500, Washington, DC 20036; 202-429-5609), formerly the Center for Environmental Education, is dedicated to protecting marine wildlife and their habitats and conserving coastal and ocean resources. The center conducts policy-oriented research, public awareness through education, and supports international and domestic marine conservation programs. It sponsors the **Whale Protection Fund**, **Marine Habitat Program**, **Seal Rescue Fund**, **Sea Turtle Rescue Fund**, and **Entanglement Program**. A four-page color pamphlet detailing the Balloon Alert Project is also available. Membership ($20 per year) includes the quarterly newsletter *Marine Conservation News*, legislative updates, and "Action Alerts" outlining things people can do to support marine conservation.

Center for Science in the Public Interest (1875 Connecticut Ave, NW, Washington, DC 20009; 202-332-9110) provides consumers with information in the areas of nutrition and health. Research, education, and publication efforts include nutrition advocacy, alcohol and minority-related projects, a child nutrition project, and the **Americans for Safe Food** project, which focuses on sustainable agriculture. Membership is $20 per year, to receive *Nutrition Action Health Letter*, published ten times annually.

Citizen's Clearinghouse for Hazardous Waste (P.O. Box 6806, Falls Church, VA 22040; 703-237-2249) was founded in 1981 by Love Canal victim Lois Gibbs to use the lessons learned at the Love Canal toxic-waste dump in helping grassroots groups fight for environmental justice. It currently works with over 6,000 community groups nationwide, providing technical support for environmental problems. The group publishes many publications geared toward helping communities help themselves—manuals on organizing, fundraising, waste disposal management, how to fight a proposed facility, and how to start a community recycling project. Membership is $25 per year and includes the bimonthly magazine *Everyone's Backyard*.

Clean Sites (1199 N. Fairfax St., Ste. 400, Alexandria, VA 22314; 703-683-8522) encourages hazardous-waste cleanups conducted by those responsible for the contamination and provides technical reviews and project management services at sites. Clean Sites is supported by reimbursement for services, contributions from the chemical industry, and contributions and grants from corporations, foundations, and government.

Clean Water Action (1320 18th St., NW, Washington, DC 20036; 202-457-1286) works for clean and safe water at an affordable cost, control of toxic chemicals, and the protection of natural resources. Emphasis is on pesticide safety and groundwater protection, solving the landfill crisis, and protecting endangered natural resources. Membership is $24 for individuals, $40 for organizations. Members receive the quarterly *Clean Water Action News* and monthly regional newsletters.

Clean Water Fund (1320 18th St., NW, 3rd Fl., Washington, DC 20036; 202-457-0336) aims to advance environmental and consumer protections and develop the grassroots strength of the environmental movement. The group focuses on water pollution, toxic hazards, and natural resources. Regular membership ($25) includes the quarterly newsletter *Clean Water Action News*; sustaining membership ($60) includes monthly bulletins as well as the newsletter.

Climate Institute (324 4th St., NE, Washington, DC 20002-5821; 202-547-0104) serves as a bridge between scientists and public and private decision-makers on global warming and stratospheric ozone depletion. The organization publishes many reports and proceedings. Membership is $95 and includes a quarterly newsletter, *Climate Alert.*

Coalition for Scenic Beauty (216 7th St. SE, Washington, DC 20003; 202-546-1100) is dedicated to protecting scenic resources in the U.S. and cleaning up visual pollution. The group has worked on billboard control, preservation of scenic areas, and aesthetic regulation. Membership ($20 individual, $50 organization) is open and includes a subscription to *Sign Control News*, a bimonthly newsletter.

The Coastal Society (5410 Grosvenor Ln., Ste. 110, Bethesda, MD 20814; 301-897-8616) is committed to promoting the understanding and wise use of coastal environments. It sponsors conferences, workshops, and publications. Membership ($25, $12.50 students) includes a quarterly bulletin.

Coastal States Organization (444 N. Capitol St. NW, Ste. 312, Washington, DC 20001; 202-628-9636) represents the coastal areas (the U.S. has a 95,000-mile coastline) in ocean and coastal affairs. The membership of congressional representatives from those states continually reviews and assesses coastal management practice and policy, problems, and progress throughout the country. It offers a broad-based information and data gathering network, provides information on coastal and offshore development, public access information, coastal hazards planning and management, wetlands preservation, fisheries development and management, and port and waterfront restoration. The group sponsors "Coastweeks" each fall, a three-week volunteer project to clean up the nation's coasts and beaches.

Concern, Inc. (1794 Columbia Rd. NW, Washington, DC 20009; 202-328-8160) provides environmental information for community action.

Concern's Community Outreach program promotes local and regional citizen action and encourages communication between individuals and groups working on similar issues. Its goal is to help communities find solutions to environmental problems that threaten public health and the quality of life. Booklets on pesticide use, farmland, drinking water, groundwater, and household waste are available at $3 each.

Conservation Foundation (1250 24th St. NW, Washington, DC 20037; 202-293-4800) is committed to improving the quality of the environment and to securing wise use of the earth's resources by influencing public policy on all levels. The group focuses on pollution and toxic substances, public and private land use in the U.S., and economic development in the third world. In 1985, the group formally affiliated with the World Wildlife Fund to add a strong scientific background to its activities. The group publishes a monthly newsletter for other environmental organizations, the *CF Newsletter*.

Consumer Pesticide Project (425 Mississippi St., San Francisco, CA 94105; 415-826-6314) is composed of consumer and environmental activists working to get dangerous pesticides out of food and the environment. The group is dedicated to encouraging citizen participation and focuses on encouraging supermarkets to carry fresh fruits and vegetables without dangerous pesticides. The project's *Organizing Kit: A Practical Strategy to Reduce Dangerous Pesticides in our Food and the Environment* is a step-by-step instruction guide for organizing your community and is available for $5 to cover postage and handling.

Co-op America (2100 M St. NW, Ste. 310, Washington, DC 20036; 202-872-5307) is a democratically controlled, nonprofit membership association representing the social and environmental interests of consumers. It believes that to solve the environmental crisis, companies must change the way they do business. The group's travel service, Travelinks, arranges ecologically friendly, educational travel. Other benefits include insurance and a credit union that offers a special Visa card. Membership ($20 per year) includes the quarterly *Building Economic Alternatives*.

Council on Economic Priorities (30 Irving Pl., New York, NY 10003; 212-420-1133) is a nonprofit research group devoted to impartial analysis of public-interest issues, including corporate social responsibility, the environment, and national security. Membership ($25 a year and up) includes a copy of the guide *Shopping for a Better World* and a monthly research report.

Cousteau Society (930 W. 21st St., Norfolk, VA 23517; 804-627-1144) was founded by world-renowned environmentalist and underwater explorer Jacques-Yves Cousteau to protect and improve the quality of life for present and future generations. The society's activities range from research, lectures, books, and publications to television specials on human interaction with ecosystems. Membership ($20 for individuals, $28 for families) includes the monthly magazine *Calypso Log* (children receive *Dolphin Log*).

Cultural Survival (11 Divinity Ave., Cambridge, MA 02138; 617-495-2562) is a Harvard-affiliated academic organization working to import sustainably managed rain forest products to the U.S. It acts as a consultant to businesses that want to import rain forest nuts and woods for use in their products. Cultural Survival manages "Cultural Survival Imports," a nonprofit business importing cashews and Brazil nuts to the U.S. Membership ($25 annually) includes a subscription to *Cultural Survival Quarterly*.

Defenders of Wildlife (1244 19th St. NW, Washington, DC 20036; 202-659-9510) is dedicated to protecting wild animals and plants in their natural communities, especially native American endangered or threatened species, through education, litigation, and advocacy of public policies. The group sponsors the Entanglement Network Coalition to identify problems with marine entanglement and debris ingestion. Membership ($20 per year) includes the bimonthly magazine *Defenders*, voting privileges for the Board of Directors, and eligibility for the organization's Visa or MasterCard. It also publishes annual endangered species reports, educational newsletters, and citizen action alerts.

Desert Fishes Council (407 W. Line St., Bishop, CA 93514; 619-872-1171) is concerned with the integrity of aquatic ecosystems in the desert Southwest. It supports the research of related agencies and academia. Membership ($10 per year) includes the proceedings of the annual symposium.

Desert Tortoise Council (5319 Cerritos Ave., Long Beach, CA 90805; 213-422-6172) was established to assure continued survival of the desert tortoise population. The group advises desert tortoise preservation agencies, fish and wildlife agencies, and any other agencies involved in protection of the desert tortoise and conservation. Membership ($8 individual, $5 student, $20 contributor, $25 organization) is open and includes a quarterly newsletter and notices of symposiums.

Ducks Unlimited (1 Waterfowl Way, Long Grove, IL 60047; 312-438-4300) raises money for developing, preserving, restoring, and maintaining the waterfowl habitat in North America. The group promotes public education about wetlands and waterfowl management, and supports the North American Waterfowl Management Plan. Membership is $20 to $10,000 and entitles members to a subscription to the monthly magazine *Ducks Unlimited*.

Earth First! (P.O. Box 7, Canton, NY 13617; 315-379-9940) is a radical direct-action movement encouraging individuals to act upon their environmental concerns. The *Earth First! Journal*, published eight times a year, is available for $20 per year.

Earth Island Institute (300 Broadway, Ste. 28, San Francisco, CA 94133; 415-788-3666) was established to initiate and support internationally oriented projects protecting and restoring the environment. Among Earth Island's projects are the International Marine Mammal Project, the Environmental Project on Central America, the International Rivers Network, and the Climate Protection Network. Membership is $25 annually and includes the quarterly *Earth Island Journal.*

EarthSave (P.O. Box 949, Felton, CA 95018; 408-423-4069) developed from the work of John Robbins, author of *Diet for a New America.* It provides education and leadership for transition to more healthful and environmentally sound food choices, nonpolluting energy supplies, and a wiser use of natural resources, and it is dedicated to an ecologically sustainable future. Membership ($20 to $35 per year) includes the *Project EarthSave* newsletter and regular notices and updates of current activities.

Elsa Wild Animal Appeal (a.k.a. Elsa Clubs of America, P.O. Box 4572, N. Hollywood, CA 90607; 818-761-8387) is dedicated to the conservation of wildlife, especially endangered species, through child education. It develops and distributes child education materials (for youth under 18) and is involved in legislation. Membership ($15 for adults, $7.50 for youth, $5 for senior citizens) includes a triennial subscription to *Born Free News* and includes a choice of several wildlife kits that can be used at home or in classrooms. Youth members also receive "Action Alerts" throughout the year with titles such as "Elephant Ivory Bans" and "Save the Whales," with suggestions and information about what they can do.

Energy Conservation Coalition (see **Environmental Action**)

Environmental Action Coalition (625 Broadway, New York, NY 10012; 212-677-1601) promotes recycling in New York City, organizing apartment buildings and urban forestry and educational programs. Membership ($15 to $25 annually) includes the quarterly newsletter *Cycle.*

Environmental Action, Inc. (1525 New Hampshire Ave. NW, Washington, DC 20036; 202-745-4870) is a membership-based organization that lobbies Congress for passage of strong environmental laws, such as the Clean Air Act and Superfund. The group works directly with citizen groups on such issues as recycling, right-to-know laws, and toxic pollution. It also sponsors the Solid Waste Alternative Project. The Energy Conservation Coalition branch works through national public interest groups to promote energy efficiency as a solution to such problems as global warming. *Environmental Action Magazine*, published bimonthly, provides in-depth articles on such topics as solid waste, plastic containers, and ecotourism. Membership ($20 per year) includes the magazine.

Environmental and Energy Study Institute (122 C St. NW, Ste. 700, Washington, DC 20001; 202-628-1400) is a public policy organization aimed at producing more informed congressional debate, credible analysis, and innovative policies for environmentally sustainable development. Programs and projects focus on global climate change, acid rain, groundwater protection, agriculture, solid and hazardous waste management, energy efficiency, and natural resource management in the third world. The group publishes special reports and the *Weekly Bulletin*, published while Congress is in session, which contains highlights of the upcoming week's floor activity in Congress, status reports and forecasts, and positions of key players.

Environmental Defense Fund (257 Park Ave. S., New York, NY 10010; 212-505-2100) was founded in 1967 as an organization of scientists, economists, and lawyers defending the environment. It focuses on water pollution, pesticides, wildlife preservation, wetland protection, rain forests, the ozone layer, acid rain, and toxic chemicals and waste. Membership is $20 per year and includes the quarterly *EDF Newsletter*.

Environmental Hazards Management Institute (EHMI, 10 Newmarket Rd., P.O. Box 932, Durham, NH 03824; 603-868-1496) aims to educate public and private sector individuals and organizations on hazardous waste, acting as an information and training source. EHMI distributes the "Household Hazardous Waste Wheel," the "Water Sense Wheel," and the "Recycling Wheel." They are available for $3.75 each plus postage and handling, with discounts for quantity orders.

Environmental Law Institute (1616 P St. NW, Ste. 200, Washington, DC 20036; 202-328-5150) is an environmental law research and education center that helps find creative solutions to such problems as wetlands protection, surface mining, hazardous waste, acid rain, and global warming. The group focuses its efforts on education through courses, publications, technical assistance, and research and policy analysis.

Environmental Safety (733 15th St. NW, #1120, Washington, DC 20005; 202-628-0374) is an association of environmental professionals who develop new approaches and alternatives for environmental policy, especially environmental safety. It monitors the activities of the EPA, including budget, personnel, and policy matters, and researches alternatives to EPA policy. It also monitors implementation of the nation's toxic-substance laws and provides information and education materials to the public.

The Forest Trust (P.O. Box 9238, Santa Fe, NM 87504; 505-983-8992) was established to protect and improve forest ecosystems and resources. The group's activities focus on national forest management, land trust, management of private lands, and economic development in rural communities.

Freshwater Foundation (2500 Shadywood Rd., Box 90, Navarre, MN 55392; 612-471-8407) is dedicated to researching and educating to keep waters usable for human consumption, industry, and recreation. Activities emphasize agricultural chemicals and groundwater protection, and biological processes that degrade pollutants. It publishes *Health and Environment Digest*, *U.S. Water News*, both monthlies, and the biennial *Journal of Freshwater*. Membership is $50 per year.

Friends of Animals (P.O. Box 1244, Norwalk, CT 06856; 203-866-5223) is dedicated to eliminating human brutality to animals. Its many programs include breeding-control services, working for the protection of animals used in experiments and testing and for farm animals, and a wild animal orphanage and rehabilitation center in Liberia. It also heads the Committee for Humane Legislation, which is active in legislative affairs. Membership is $20 per year.

Friends of the Earth (530 7th St. SE, Washington, DC 20003; 202-544-2600) promotes the conservation, protection, and rational use of the earth. Its activities include lobbying, litigation, and public information on a variety of environmental issues, including ozone depletion, river protection, and tropical deforestation. It recently merged with the Oceanic Society and continues to strive to protect the oceans through education, research, and conservation, and by promoting the understanding and stewardship of the marine and coastal environment. Membership ($25 individual, $15 student/low income/senior citizen) includes the monthly magazine *Not Man Apart*.

Friends of the River (Fort Mason Center, Bldg. C, San Francisco, CA 94123; 415-771-0400) was established to save rivers from being dammed. It is active in river conservation efforts, and has successfully fought for preservation of the Tuolumne, Kings, Kern, and Merced rivers in California. Membership is $25 for individuals, $35 for families; members receive the bimonthly *Headwaters* as well as "action alerts" and discounts on river trips.

Friends of the Sea Otter (Box 221220, Carmel, CA 93922; 408-625-3290) aids in the protection and maintenance of the southern sea otter and its nearshore marine habitats. Membership is $15 per year and includes the twice-annual *Otter Raft* magazine.

Fund for Animals (200 W. 57th St., New York, NY 10019; 212-246-2096) aims to aid the relief of fear, pain, and suffering in wild and domestic animals.

Annual membership is $10 for students, $20 for individuals, and $25 for families, including a newsletter and "Action Alert" updates on legislation.

Global Tomorrow Coalition (1325 G St. NW, Washington, DC 20005; 202-628-4016) is a national alliance of organizations and individuals created to foster broader understanding of the long-term significance of global trends in population, resources, environment, and development. Its main focus is on sustainable development. The group also helps promote informed and responsible public choice among alternative futures for the U.S. and alternative roles for the nation within the international community. Membership is $15 a year for senior citizens and students and $35 for individuals and includes the quarterly newsletter *InterAction*.

Grass Roots the Organic Way (38 Llangollen Lane, Newton Square, PA 19073; 215-353-2838) provides information about harmful pesticides and safe alternatives. It educates on the use and misuse of pesticides on lawns, trees, shrubs, and other indoor and outdoor plants. Membership ($20) entitles members to call with specific problems. The group will either provide solutions over the phone or will send information on the subject.

Great Bear Foundation (P.O. Box 2699, Missoula, MT 59806; 406-721-3009) is dedicated to conservation of bears, especially grizzly bears, and their habitat through such means as public education and habitat monitoring. Membership is $15 for individuals, $25 for families, and includes the quarterly newsletter *Bear News*.

Great Swamp Research Institute (Office of the Dean, College of Natural Sciences and Math, Indiana University of Pennsylvania, 305 Weyandt Hall, Indiana, PA 15705; 412-357-2609) is a research and educational organization committed to protecting, preserving, and maintaining the environment. It seeks new solutions to deal with the daily pressures society is placing on the environment. Membership is free; write for information.

Greater Yellowstone Coalition (P.O. Box 1874, Bozeman, MT 59715; 406-586-1593) aims to preserve and protect the wildlife, wildlands, fisheries, and other natural wonders in and around Yellowstone National Park. Membership is $20 for individuals, $50 for organizations, per year.

Greenhouse Crisis Foundation (1130 17th St. NW, Ste. 630, Washington, DC 20036; 202-466-2823), a project of the Foundation on Economic Trends, is dedicated to creating global awareness of the greenhouse crisis and to changing the world view and lifestyle underlying that crisis. It has initiated several local, national, and international actions, including the Greenhouse Education Campaign; the Cities Program to raise awareness of mayors throughout the world of the importance of energy conservation, urban reforestation, and mass transit; International Environmental Rights Conferences; and the Global Greenhouse Network, to raise awareness and mobilize public opinion and to help facilitate international cooperation.

Greenpeace (1436 U St. NW, Washington, DC 20009; 202-462-1177) is dedicated to protecting and preserving the environment and the life it supports. It has focused its efforts on halting the needless killing of marine mammals and other endangered animals, ocean ecology, toxic waste reduction, and nuclear disarmament. Membership ($20 per year) includes the bimonthly *Greenpeace* magazine.

Household Hazardous Waste Project (901 S. National Ave., Box 108, Springfield, MO 65804; 417-836-5777) educates the public on household hazardous waste. It provides consumer information, offers training and materials to community groups, and supports a grassroots approach to working on household hazardous waste issues.

Human Ecology Action League (P.O. Box 49126, Atlanta, GA 30359; 404-248-1898) aims to increase awareness of environmental conditions that are hazardous to human health. The group acts as a clearinghouse on chemical sensitivities and related disorders, works toward minimizing the indiscriminate use of harmful chemicals, and establishes local chapters that provide support for members and educate their communities. Membership ($20) includes a subscription to the newsletter *The Human Ecologist.*

Human Environment Center (1001 Connecticut Ave. NW, Ste. 827, Washington, DC 20006; 202-331-8387) is dedicated to providing education, information, and services to encourage the integration of environmental organizations, and promoting joint activities among environmental and social equity groups. It serves as a clearinghouse and technical assistance center for youth conservation and service corps programs and operates a recruitment and placement service for minority environmental interns and professionals.

Humane Society of the United States (2100 L St. NW, Washington, DC 20037; 202-452-1100) offers resources to the general public on such topics as animal control, cruelty investigation, publications, and humane education. The group's efforts include a "Shame of Fur" campaign, "Be a P.A.L.—Prevent A Litter" campaign, marine mammal protection, and laboratory animal welfare. Membership is $10 per year.

Infact (256 Hanover, Boston, MA 02113; 617-742-4583) was established to press for corporate accountability and responsibility. The group has research offices in several cities and national campaign headquarters in California. One focus is on protesting the nuclear weapons efforts of

General Electric. The publication *Infact Brings GE to Light* is available for $8.45. Membership is $15 a year.

Inform (381 Park Ave. S., New York, NY 10016; 212-689-4040) was established to conduct environmental research and education on such topics as garbage management, industrial toxic waste reduction, urban air pollution, and land and water conservation. Membership ($35 and up) includes a bimonthly newsletter.

Institute for Earth Education (Box 288, Warrenville, IL 60555; 312-393-3096) is dedicated to developing a serious educational response to the environmental crisis of the earth. It consists of a network of individuals and member organizations committed to fostering earth education programs. The group conducts workshops, publishes a quarterly journal, *Talking Leaves*, hosts international and regional conferences, supports local branches, and publishes books. Membership ($20 and up) includes a subscription to the journal. A free *Sourcebook* outlining the group's Earth Education program materials is available upon request.

Institute for Local Self-Reliance (2425 18th St. NW, Washington, DC 20009; 202-232-4108) helps communities achieve maximum use of their physical, financial, and human resources. Its activities include research, educational workshops, and direct consulting services to community groups and local governments.

International Alliance for Sustainable Agriculture (Newman Center, University of Minnesota, 1701 University Ave. SE, Rm. 202, Minneapolis, MN 55414; 612-331-1099) works to promote sustainable agriculture worldwide in an economically viable, ecologically sound, socially just, and humane way. The group focuses on three problems in sustainable agriculture: insufficient research and documentation; lack of organizational support and network building; and inadequate education and information dissemination. Membership ($10 to $1,000) includes a subscription to *Manna*, a bimonthly newsletter, and discounts on other publications.

International Council for Bird Preservation (801 Pennsylvania Ave. SE, Washington, DC 20003; 202-778-9563) was established to help maintain the diversity, distribution, abundance, and natural habitats of bird species worldwide, and to prevent the extinction of any species or subspecies. Activities have focused on monitoring the status of susceptible bird populations and fostering international cooperation in bird preservation efforts. Membership ($35 and up) entitles members to two quarterly magazines, *World Bird Watch* and *U.S. Bird News*.

International Fund for Animal Welfare (P.O. Box 193, 411 Main St.,

Yarmouth Port, MA 02675; 508-362-4944) is dedicated to the protection of wild and domestic animals. Efforts have included work to preserve harp and hood seals in Canada, fur seals in Alaska, and vicuna in Peru. Membership is by donation; benefits include a variety of newsletters.

International Oceanographic Foundation (4600 Rickenbacker Causeway, P.O. Box 499900, Miami, FL 33149; 305-361-4888) provides information about the world's oceans and their importance to humanity, and encourages scientific investigation of the ocean. It operates a museum at its Miami headquarters. Membership is $18 and entitles members to receive the bimonthly magazine *Sea Frontiers*.

International Primate Protection League (P.O. Box 766, Sumerville, SC 28484; 803-871-2280) is dedicated to the conservation and protection of primates. Activities focus on primate trafficking, laboratory primate issues, and maintenance of a gibbon sanctuary. Membership is $20 and includes the quarterly newsletter *International Primate Protection League*.

International Society of Arboriculture (303 W. University Ave., Urbana, IL 61801; 217-328-2032) is dedicated to proper tree care and preservation, particularly in urban settings. Its activities are aimed at helping make the public aware of the impact trees have on our future. The group provides services to its members about the science and art of growing and maintaining shade and landscape trees. Membership ($55 per year) includes a subscription to the monthly *Journal of Arboriculture*.

International Society of Tropical Foresters (5400 Grosvenor Lane, Bethesda, MD 20814; 301-897-8720) is dedicated to protecting, wisely managing, and rationally using the world's tropical forests. Activities include establishing a communications network among tropical foresters and others concerned with the forests. Membership ($50 annually) includes a subscription to a quarterly newsletter *ISTF News*, and the monthly magazine *Journal of Forestry*.

Izaak Walton League of America (1701 N. Fort Myer Dr., Arlington, VA 22209; 703-528-1818) aims to protect America's land, water, and air resources. Efforts focus on acid rain, clean air and water, stream protection, soil erosion, the Chesapeake Bay cleanup, and waterfowl/wildlife protection. The organization coordinates the programs Save Our Streams and Wetlands Watch. Membership ($20 per year) includes a quarterly magazine *Outdoor America*.

League of Conservation Voters (320 4th St. NE, Washington, DC 20002; 202-785-8683) is a nonpartisan political arm of the environmental movement. It works to elect pro-environmental candidates to Congress, based on energy, environment, and natural-resource issues. Membership is $25 per year and includes *The National Environmental Scorecard*, an annual rating of members of Congress on environmental issues.

League of Women Voters of the United States (1730 M St. NW, Washington, DC 20036; 202-429-1965) is a nonpartisan, political organization that encourages informed, active participation of citizens in government and influences public policy through education and advocacy. The league takes political action on water and air quality, solid- and hazardous-waste management, land use, and energy. Membership is $50 per year for a national membership (local memberships vary) and includes a subscription to the monthly *National Voter* magazine.

Marine Mammal Stranding Center (P.O. Box 733, Brigantine, NJ 08203; 609-266-0538) is a rescue and rehabilitation organization for marine mammals and sea turtles. It sponsors many whale-, dolphin-, and seal-watching trips. Membership is $10 per year and includes the newsletter *Blow Hole*.

Monitor Consortium of Conservation and Animal Welfare Organizations (1506 19th St. NW, Washington, DC 20036; 202-234-6576) was founded in 1972 as a nonprofit coordinating center and information clearinghouse on endangered species and marine mammals for its member organizations. Membership is open to conservation, environmental, and animal welfare organizations.

National Arbor Day Foundation (100 Arbor Ave., Nebraska City, NE 68410; 402-474-5655) is dedicated to tree planting and conservation. It provides direction, technical assistance, and public recognition for urban and community forestry programs. Projects include Tree City USA, an urban forestry tree planting and care program; Friends of Tree City, for individuals in urban areas that are interested in trees; Trees for America; and Conservation Trees. The group provides materials and information for cities to plan their own Arbor Day celebrations. Membership ($10 per year) includes the bimonthly newsletter *Arbor Day* and a tree book.

National Audubon Society (950 Third Ave., New York, NY 10022; 212-832-3200) aims to conserve native plants and animals and their habitats; protect life from pollution, radiation, and toxic substances; further the wise use of land and water; seek solutions for global problems involving the interaction of population, resources, and the environment; and promote rational strategies for renewable energy development. Membership ($30 per year) includes a subscription to the bimonthly *Audubon Magazine*. A major activity is the 12,000-member Activist Network. Network members volunteer to act upon their environmental concerns, especially in times of environmental crisis, such as an oil spill. Membership in the network ($9 per year) includes a subscription to the bimonthly *Audubon Activist* and *Action Alerts*, reporting the status of environmental legislation. The society also offers environmentally responsible travel programs, school programs, adult conservation camps, and an ornithological journal, *American Birds*.

National Clean Air Coalition (530 7th St. SE, Washington, DC 20003; 202-543-8200) is a lobbying and education coalition that aims to address clean-air issues. Membership is free and includes a periodical newsletter, *Clean Air 101*.

National Coalition Against the Misuse of Pesticides (530 7th St. SE, Washington, DC 20003; 202-543-5450) assists individuals, organizations, and communities with information on pesticides and their alternatives. The group maintains an information clearinghouse that provides materials on agricultural and urban issues concerning lawn-care safety, farm workers' safety, groundwater problems, and alternatives to pesticides. Membership ($20) includes the newsletter *Pesticides and You*.

National Coalition for Marine Conservation (P.O. Box 23298, Savannah, GA 31403; 912-234-8062) is dedicated to conserving oceanic ecosystems and the habitat areas that support them. It works to educate policymakers at all levels of government. Efforts focus on fishery management, oceanic dumping, and wetlands preservation. Membership ($25 annually) includes a bimonthly newsletter *Marine Bulletin*.

National Geographic Society (17th and M Sts. NW, Washington, DC 20036; 202-857-7000), founded in 1888 "for the increase and diffusion of geographic knowledge," is now the world's largest scientific and educational nonprofit organization, with 10.8 million members. Its main areas of activity are publishing four magazines, books for adults and children, and atlases; producing television programs; supporting environmental research and a geography education program for schoolchildren. Membership is $21 a year and includes a subscription to *National Geographic* magazine.

National Institute for Urban Wildlife (10921 Trotting Ridge Way, Columbia, MD 21044; 301-596-3311) is devoted to wildlife research, management, and conservation education programs and activities. Projects have included discovering practical procedures for maintaining or enhancing wildlife species in urban areas. Membership ($25) includes the quarterly newsletter *Urban Wildlife News*.

National Parks and Conservation Association (1015 31st St. NW, Washington, DC 20007; 202-944-8530) is dedicated to defending, promoting, and improving America's national park system while educating the public about the parks. Membership ($25) entitles members to the bimonthly magazine *National Parks* as well as discounts on car rentals.

National Recycling Coalition (1101 30th St. NW, Ste. 305, Washington, DC 20006; 202-625-6406) promotes increased opportunities for recycling. It focuses on education, information, and lobbying, and sponsors the largest annual conference on recycling, the Annual National Recycling Congress. The group's Technical Assistance Program sends advisers to cities that want to establish recycling programs. Membership ($30 per year) includes a bimonthly newsletter *NRC Connection*.

National Toxics Campaign (37 Temple Pl., 4th Fl., Boston, MA 02111; 617-482-1477) works on Superfund-related activities. The group provides citizen outreach and educational efforts, political organizing, and a research library. It also maintains a testing lab and legal offices that provide advice to communities dealing with hazardous-waste issues. Membership is $25 for an individual and $50 and up for community groups, and includes the quarterly magazine *Toxic Times*.

The National Water Center (P.O. Box 264, Eureka Springs, AR 72632; 501-253-9755) aims to gather, distill, and disseminate information on water issues, emphasizing personal responsibility for human and hazardous waste. Activities include a focus on promoting composting toilets. Membership (minimum $10 per year) includes the newsletter *Water Center News*.

National Wildlife Federation (1400 16th St. NW, Washington, DC 20036; 202-797-6800) was founded in 1936 "to be the most responsible and effective conservation education association promoting the wise use of natural resources and protection of the global environment." It distributes periodicals and educational materials, sponsors education programs in conservation, and litigates environmental disputes. Efforts have focused on forests, energy, toxic pollution, environmental quality, biotechnical fisheries and wildlife, wetlands, water resources, and public lands. Membership rates vary depending on magazine subscriptions desired: *National Wildlife* and *International Wildlife*, monthly magazines, are $15 each per year; *Ranger Rick*, a monthly magazine for children 6 to 12 years old, is $14 per year; *Your Big Backyard*, a monthly magazine for children aged 3 to 5 years old, is $10.

Natural Resources Defense Council (40 W. 20th St., New York, NY 10011; 212-727-2700) aims to protect America's endangered natural resources and to improve the quality of the environment. It monitors government agencies, brings legal actions, and disseminates information. Areas of focus include air and water pollution, global warming, urban environment, toxic substance, resource management, energy conservation, and Alaska. Membership ($10) includes a subscription to *The Amicus Journal* (quarterly) and the *Natural Resources Defense Council Newsline*, a bimonthly newsletter.

The Nature Conservancy (1815 N. Lynn St., Arlington, VA 22209; 703-841-5300) acts to preserve ecosystems and the rare species and communities they shelter. The group has protected more than 3.5 million acres of threatened habitat, mostly by purchasing land, and manages more than 1,000 preserves. It maintains a database, the Heritage Network, which is a national inventory of species in each state. Membership is $15 and includes a bimonthly magazine *The Nature Conservancy Magazine*, as well as newsletters and update information from state chapters.

North American Lake Management Society (P.O. Box 217, Merrifield, VA 22116; 202-466-8550) is dedicated to promoting a better understanding of lakes, ponds, reservoirs, impoundments, and their watersheds. The group encourages the exchange of information about lake management; provides guidance to public and private agencies involved in lake management; and promotes research on lake ecology and watershed management. Membership ($20 and up) includes a periodic newsletter *Lakeline*.

Oceanic Society (see **Friends of the Earth**)

Pacific Whale Foundation (101 N. Kihei Rd., Ste. 21, Kihei, Maui, HI 96753; 808-879-8811, 800-942-5311) is dedicated to the science of saving whales and the ocean environment. It conducts public education programs on a variety of cetacean, marine conservation, and pollution issues, and it sponsors a Marine Debris Cleanup Day and Whale Day. Membership is $15 for students and seniors, $20 for individuals, and $25 for families.

Pennsylvania Resources Council (25 W. 3rd St., P.O. Box 88, Media, PA 19063; 215-565-9131) sponsors a statewide recycling conference every spring and publishes *Environmental Shopping*, a booklet designed to help shoppers choose environmentally safe products. The group publishes two newsletters, *PRC News* (3 times per year) and *All About Recycling* (quarterly). Newsletters are free with membership ($30 for individuals; $35 for nonprofit organizations; $50 and up for businesses).

People for the Ethical Treatment of Animals (P.O. Box 42516, Washington, DC 20015; 301-770-7444) aims to end the exploitation and abuse of animals, with a special focus on animals used in laboratory experiments. The group sponsors an anti-fur campaign and a vegetarian campaign as well as lobbying, public education efforts, and demonstrations for animal rights. Membership ($15 per year) includes the bimonthly news magazine *PETA News*.

The Peregrine Fund (5666 W. Flying Hawk Lane, Boise, ID 83709; 208-362-3716) is devoted to preserving rare and endangered birds of prey worldwide. It coordinates field projects to study birds of prey and conducts reintroduction programs and educational programs. Projects include the Peregrine Falcon Recovery, Aplomado Falcon Recovery, and tropical raptor surveys. Membership is $25 a year and includes *The Peregrine Fund Newsletter*.

Pesticide Action Network (P.O. Box 610, San Francisco, CA 94101; 415-541-9140) is a worldwide coalition of more than 300 organizations in more than 50 countries working to stop pesticide misuse and global pesticide proliferation and work toward safe, sustainable pest control. The group produces two triennial newsletters, the *Global Pesticide Monitor* and the *Dirty Dozen Campaigner*, the bimonthly *PANNA Outlook*, and publications on pesticides and alternatives. Membership ranges from $30 to $50 and includes a subscription to *Global Pesticide Monitor*.

Public Citizen (2000 P St. NW, Washington, DC 20036; 202-293-9142) was established by Ralph Nader in the interest of consumer protection. It conducts research, lobbying, lawsuits, expert testimony, and publications on issues including stronger environmental programs and safe energy. Membership ($20 per year) includes a subscription to the monthly *Public Citizen Magazine*.

Public Voice for Food and Health Policy (1001 Connecticut Ave. NW, Ste. 522, Washington, DC 20036; 202-659-5930) is a consumer research, education, and advocacy organization working on issues related to health, nutrition, and food safety. Special emphasis is placed on establishing comprehensive, mandatory seafood testing; access to a safe, affordable, and nutritious food supply; and women's health issues. Membership is $20 per year and includes the quarterly *Action Alert* bulletins highlighting recent legislation.

Rachel Carson Council (8940 Jones Mill Rd., Chevy Chase, MD 20815; 301-652-1877) is an international clearinghouse of information on ecology of the environment for all on chemical contamination. It sponsors conferences and produces publications. Subscription to its newsletter is $15 per year and includes Council Publications for one year.

Rainforest Action Movement (430 E. University, Ann Arbor, MI 48109; 313-764-2147) focuses on the preservation, protection, and rational use of rain forests in Alaska, Oregon, Washington, Hawaii, and tropical areas. It sponsors public awareness programs in Ann Arbor and surrounding areas, including programs for schoolchildren, lectures, films, citizen forums, benefit concerts, and weekly activities on the campus. Its newsletter *Tropical Echoes*, published every six weeks, is $10 a year.

Rainforest Action Network (300 Broadway, Ste. 28, San Francisco, CA 94133; 415-398-2732) focuses exclusively on rain forest protection. It works with other environmental and human rights organizations on major rain forest protection campaigns. Topics of emphasis include tropical timber, indigenous peoples, and multilateral development banks. The group has educational publications and materials, a slide show, and press kits available. Membership ($25 regular, $15 limited income) includes two newsletters, *Rain Forest Action Alert* (monthly) and *World Rain Forest Report* (quarterly).

Rainforest Alliance (270 Lafayette St., Ste. 512, New York, NY 10012; 212-941-1900) aims to link individuals interested in saving tropical rain forests. The group brings together conservation groups, professional organizations, financial institutions, scientists, the business community, and concerned individuals. Membership is $20 per year, $15 for students and senior citizens, and includes a subscription to the quarterly newsletter *Canopy*.

Renew America (1001 Connecticut Ave. NW, Ste. 719, Washington, DC 20036; 202-232-2252) is an educational and networking forum devoted to the efficient use of natural resources, working at the federal, state, and private citizen levels. Membership ($25) includes the quarterly *Renew America Report* and a copy of the yearly *State of the States* report, which provides an annual "report card" on current developments across the nation.

Resources for the Future (1616 P St. NW, Washington, DC 20036; 202-328-5000) conducts research on the environment and the conservation and development of natural resources, including air and water pollution, solid waste disposal, pesticides, toxic substances, and international issues. It publishes various research and technical publications. Although it is not a membership organization, a quarterly newsletter, *Resources*, is available free.

Rocky Mountain Institute (1739 Snowmass Creek Rd., Old Snowmass, CO 81654; 303-927-3128) aims to foster the efficient and sustainable use of resources as a path to global security. It offers its research on resource efficiency, global security, and community economic renewal through its publications. A publications list is available upon request.

Save Our Streams (see **Izaak Walton League of America**)

Save the Redwoods League (114 Sansome St., Rm. 605, San Francisco, CA 94104; 415-362-2352) was established to rescue areas of primeval forest from destruction. The league purchases redwood groves by private subscription and encourages a better, more general understanding of the value of primeval forests. Projects include protecting giant sequoias at Sequoia National Forest, working to establish the Smith Wild and Scenic National Park in northern California, and purchasing additional redwood lands. Membership ($10 and up) includes spring and fall bulletins on the league's activities.

Sea Shepherd Conservation Society (P.O. Box 7000-S, Redondo Beach, CA 90277; 213-373-6979) is an all-volunteer organization established to protect marine animals and marine habitats by direct action. Activities include preventing the killing of dolphins by the tuna industry, protecting pilot whales in the Faroe Islands, and rescuing whales and marine mammals in distress. Membership is by donation. Members receive the quarterly newsletter *Sea Shepherd Log*.

Sierra Club (730 Polk St., San Francisco, CA 94109; 415-776-2211) promotes conservation of the natural environment by influencing public policy decisions. It was founded in 1892 to explore, enjoy, and protect the wild places on earth; to practice and promote the responsible use of the earth's ecosystems and resources; and to educate and enlist humanity to protect and restore the quality of the natural and human environment. Campaigns include the Clean Air Act reauthorization, arctic national wildlife refuge protection, national parks and forest protection, global warming/greenhouse effect, and international development lending reform. Membership ($33 per year) includes a subscription to the monthly magazine *Sierra* and chapter publications. The club also offers more than 275 outings annually. The group's **Legal Defense Fund** (2044 Fillmore St., San Francisco, CA 94115; 415-567-6100) supports lawsuits brought on behalf of citizens' organizations to protect the environment.

Soil and Water Conservation Society of America (7515 NE Ankeny Rd., Ankeny, IA 50021; 515-289-2331) is a scientific and educational organization dedicated to conservation of land, water, and other natural resources. Efforts focus on low-input sustainable agriculture, water quality, land and water conservation, and evaluation of farm legislation. The society sponsors a scholarship program for high school and college students. Membership (first year $25, $37 thereafter) includes the *Journal of Soil and Water Conservation*, a bimonthly, and a newsletter covering the annual meeting.

Student Conservation Association (P.O. Box 550, Charlestown, NH 03603; 603-826-4301) provides high school and college students and others with the opportunity to volunteer their services for the better management of national parks, forests, public lands, and natural resources. Membership ($10 for students, $25 for others) includes the annual programs listing and the newsletter *The Volunteer*.

Student Pugwash (1638 R St. NW, Ste. 32, Washington, DC 20009; 202-328-6555) sponsors a variety of educational programs for university students, preparing them as "future leaders and concerned citizens to make thoughtful decisions about the use of technology." Student Pugwash runs international conferences, a "New Careers" program, has chapters on 30 campuses nationwide, and distributes educational products and publications. The group produces several publications, including two newsletters: *Pugwatch*, distributed to university students, and *Tough Questions*.

TreePeople (12601 Mulholland Dr., Beverly Hills, CA 90210; 818-753-4600) is dedicated to promoting personal involvement, community action, and global awareness of environmental issues. Efforts include teaching people how to plant and maintain trees, environmental leadership programs for children, and reforestation efforts in the mountains surrounding Los Angeles. Membership ($25 and up) includes the bimonthly newsletter *Seedling News* and six free seedlings each year.

Trees for Life (1103 Jefferson, Wichita, KS 67203; 316-263-7294) provides funding, management, and know-how to people in developing countries to plant and care for food-bearing trees. It runs the "Grow-a-Tree" program, encouraging children to plant trees, and distributes packets of materials, seeds, and instructions to schools and summer camps. Membership (free) includes a subscription to the quarterly newsletter *Life Lines*.

Trust for Public Land (116 New Montgomery St., San Francisco, CA 94105; 415-495-4014) aims to conserve land as a living resource for present and future generations. The trust has helped or established more than 150 local land trusts and has acted to preserve nearly 500,000 acres under public ownership. Activities focus on urban waterfronts, suburban greenways, wetlands, agricultural lands, and inner-city open spaces.

U.S. Public Interest Research Group (215 Pennsylvania Ave. SE, Washington, DC 20003; 202-546-9707) focuses on consumer and environmental protection, energy policy, and governmental and corporate reform. Efforts include monitoring the implementation of Superfund and legislation for clean air and pesticide safety. Membership ($25) includes the quarterly *Citizen's Agenda*.

Union of Concerned Scientists (26 Church St., Cambridge, MA 02138; 617-547-5552), perhaps best known for its "doomsday clock," is a coalition of scientists, engineers, and other professionals concerned with health, safety, environmental, and national security problems posed by nuclear energy and weapons. It conducts policy and technical research, public education, and legislative advocacy on advanced-technology issues.

Wetlands Watch (see **Izaak Walton League of America**)

The Whale Center (3929 Piedmont Ave., Oakland, CA 94611; 415-654-6621) promotes whales and their ocean habitats through conservation, education, research, and advocacy. Activities have centered on a whaling moratorium, establishment of national marine sanctuaries, the "WhaleBus" mobile classroom, and the "Adopt-a-Gray-Whale" program. Annual membership is $25 for individuals, $10 for students and seniors, $35 for families, and $250 for "life" members. Members receive the center's quarterly newsletter *The Whale Center Journal.*

The Wilderness Society (1400 Eye St. NW, Washington, DC 20005; 202-842-3400) aims to protect wildlands, wildlife, forests, parks, rivers, and shorelands. Efforts focus on an Arctic wildlife refuge, national park and ecosystem management, and a national forest policy. Membership ($15 the first year, $30 subsequent years) includes a subscription to the bimonthly newsletter *The Wildlifer.*

Wildlife Conservation International (New York Zoological Society, Bronx, NY 10460; 212-367-1010) is an international conservation program. It conducts field research and conservation action geared toward obtaining a more complete understanding of the biology of endangered species and the structure, functioning, and stability of large ecosystems. Membership ($23 associates, $50 supporting) includes a subscription to the *Wildlife Conservation International Newsletter.*

Wildlife Information Center (629 Green St., Allentown, PA 18102; 215-434-1637) is dedicated to securing and disseminating wildlife conservation, recreation, and scientific research information. Its programs include in-service teacher training courses and other public education, maintaining a Wildlife Conservation Registry of Fame, sponsoring wildlife conferences, and advocating nonlethal uses of wildlife such as observation, photography, sound recording, drawing and painting, and wildlife tourism. Projects include securing a ban on importation and sale of live wild birds as pets and banning the use of pole traps. Membership ($25 and up) includes a subscription to *Wildlife Activist.*

Wildlife Society (5410 Grosvenor Lane, Bethesda, MD 20814; 301-897-9770) is active in scientific management of the earth's wildlife resources management. Membership ($20 individual, $10 students) includes a subscription to *The Journal of Wildlife Management*, *The Wildlifer*, and *Wildlife Society Bulletin.*

Windstar Foundation (2317 Snowmass Creek Rd., Snowmass, CO 81654; 303-927-4777) is an educational organization dedicated to the belief that responsible personal action is the key to creating a sustainable future on a global scale. It conducts research, develops demonstration projects, and offers educational programs. Membership ($35 per year) includes the quarterly *Windstar Journal.*

World Environment Center (419 Park Ave. S., New York, NY 10016;

212-683-4700) serves as a bridge between industry and government to strengthen environmental management and industrial safety through exchange of information and technical expertise. The center publishes *The World Environment Handbook*.

World Resources Institute (1735 New York Ave. NW, Washington, DC 20006; 202-638-6300) helps the government, the private sector, and organizations address issues in environmental integrity, resources management, economic growth, and international security. Each year it publishes the *World Resources Report*.

Worldwatch Institute (1776 Massachusetts Ave. NW, Washington, DC 20036; 202-452-1999) is an independent research organization alerting decision-makers and the general public to emerging global trends in the availability and management of resources. The results of its research are published in Worldwatch papers and books. It also publishes the *State of the World* series, an annual report on the world's resources and their management. Worldwatch papers and books are available by subscription ($25 per year). It also publishes a bimonthly magazine *World Watch* ($20 per year).

WorldWIDE (1250 24th St. NW, Washington, DC 20037; 202-331-9863) aims to mobilize women to maintain and improve environmental quality and natural resource management, and to educate the public about the linkages among women, natural resources, and sustainable development. It sponsors WorldWIDE forums and publishes the *Directory of Women in the Environment*. Membership ($35 per year) includes a subscription to the newsletter *WorldWIDE News*.

World Wildlife Fund (1250 24th St. NW, Washington, DC 20037; 202-293-4800) works to protect endangered wildlife and wildlands. Its top priority is conservation of the tropical forests in Latin America, Asia, and Africa. The group supports individuals and institutions carrying out practical, scientifically based conservation projects. Members ($15 per year) receive *Focus*, a bimonthly newsletter and periodic letters about upcoming projects and travel programs.

The Xerces Society (10 SW Ash St., Portland, OR 97204; 503-222-2788) promotes the global protection of invertebrate habitats, and fosters positive public knowledge of insects by emphasizing their beneficial roles in natural ecosystems. Efforts have aimed at preserving monarch butterfly wintering habitat and creating a database of invertebrate specialists. Membership ($15 students/seniors/retirees, $25 others) includes a subscription to the magazine *Wings*.

INDEX

Above the Clouds Trekking, 267
acid rain, 14-16
Acid Rain Foundation, 306
A Citizen's Toxic Waste Audit Manual, 289
Acorn Designs, 279
Adopt-a-Stream Foundation, 306
Adventure Center, 267
aerosols, 38-39, 103, 168
AFM Enterprises, Inc., 252
African Wildlife Foundation, 306
agricultural chemicals, 30
agriculture, 118-126, 176
  low-impact, 121-122
air conditioners, 217-221
air conditioning, automobile, 65, 82-84
air fresheners, 168-169
air pollution, 14-16, 21-23, 55, 147, 247-248
Alaska Wildland Adventures, 267
All-One God Faith, 167
all-purpose cleaners, 169
Alliance to Save Energy, 306
All the Best Pet Care, 201
alternative fuels, 56-64
alternative trade organizations, 282-284
aluminum, 33, 45-46
Amazonia Expeditions, 267
American Association of Zoological Parks and Aquariums, 271-272
American Cave Conservation Association, 307
American Cetacean Society, 307
American Council for an Energy-Efficient Economy, 15-16, 164-165, 206, 307
American Forest Association, 307
American Forest Council, 307
American Forestry Association, 194
American Hiking Society, 272
American Humane Association, 308
American Littoral Society, 308
American Rivers, 308
Americans for Safe Food, 125-126, 309
Americans for the Environment, 309
American Sheep Industry, 983
American Society for Environmental History, 308
American Society for the Prevention of Cruelty to Animals, 308-309
American Wildlands, 309
America the Beautiful Fund, 306-307
Animal Protection Institute of America, 309
animal testing, 110-113
Animal Town, 276
Animal Welfare Institute, 309
antifreeze/coolant, 84
Apex Distributing, 246
appliances, 165, 204-227
  life-cycle costs, 207
aquariums, 271-272
ARCO, 60
aseptics, 127-128
Ashdun Industries, 155, 157
A Socially Responsible Financial Planning Guide, 301
Association for Commuter Transportation, 86, 88
A Thread of Hope, 284
Atlantic Center for the Environment, 309
Atlantic Recycled Paper Company, 279
Audubon Society, 290
Auro Organic Paints, 252
automobiles, 22-23, 55-88
  air conditioning, 82-84
  antifreeze/coolant, 84
  batteries, 55, 79
  mechanics, 73
  electric, 61-64
  fuel-boosting gadgets, 77
  fuel-efficient, 16, 68-74
  hazardous substances in, 79-84
  maintenance of, 71
  polluting impact of, 55-56
  purchasing, 64-68
  tires, 51
  used, 67

Baby Bunz & Co., 101
baby foods, 150-151
baby products, 110
Bales Furniture Manufacturing, 256
Banta, John, 249
barbecuing, 147

Basically Natural, 112
Basic Foundation, 193-194, 281, 309-310
Bat Conservation International, 310
batteries, 12, 28, 241-243
    automobile, 55, 79
    recycling, 46-47
Baubiologie Hardware, 249
beef, 26, 139-140
The Better World Investment Guide, 301
beverages, 126-137
Beyond Beef, 140, 142
Beyond Beef Campaign, 143
Bio-Integral Resource Center, 310
Biobottoms, 101
biodegradable cleaning products, 158-159, 161
biodiversity, 23-26, 131
Biological Journeys, 267
Biomatik, 103
bleach, 160, 162
bleached paper products, 153-154, 156-157
The Body Shop, 102
bottled water, 31-32, 129-133
boycotts, 294-297
Breeders Equipment Co., 202
Bronner's Pure-Castile Soap, 105
Building a Healthy Lawn, 187
Building With Junk, 254
Bumble Bee, 144
Bumkins, 102

California Certified Organic Farmers, 124
California Public Interest Research Group, 118
Campbelll Soup Company, 142
Campus Ecology, 295
carbon dioxide, 17, 24, 87
The Car Book, 67
cardboard, 47
Caretta Research Project, 269
Caribbean Conservation Corp., 310
carpet deodorizers, 169
carpooling, 85-87
cars, *see* automobiles
Carson, Rachel, 177
cattle farming, 141
Center for Environmental Information, 310
Center for Investigative Reporting, 310-311
Center for Marine Conservation, 275, 311
Center for Responsible Tourism, 265, 266
Center for Science in the Public Interest, 311
Centers for Disease Control, 120
Ceres Principles, 304-305
Chanchich Lodge, 271
Chanel, 103
Chevrolet, 56-57
Chicken of the Sea, 144
childrens' toys, 273-277
ChillShield, 209
chlorine, 160
chlorofluorocarbons, 17, 19-20, 65, 82-84, 103, 253
Christian Council of Asia, 263
Christmas trees, 193
Church and Dwight, 166-167
Citizen's Clearinghouse for Hazardous Waste, 311
citizen involvement, 287-297
clean air, *see* air pollution
cleaning products, 32, 158-174
    biodegradable, 158-159, 161
    natural, 168-174
    using safely, 173
    see also laundry detergents
Clean Sites, 311
Clean Water Action, 311-312
Clean Water Fund, 312
Climate Institute, 312
Clothcrafters, 157
cloth diapers, 98-102
clothing, 277-278
    recycling, 47-48
Co-op America, 246, 277, 279-280, 282-283, 297, 301, 313
Coalition for Environmentally Responsible Economies, 304-305
Coalition for Scenic Beauty, 312
The Coastal Society, 312
Coastal States Organization, 312
coffee, 133-134
coffee filters, 133-134, 156-157
Cohn, Susan, 305
community action, 292-293
commuting, 85-88
    pollution caused by, 59
compact fluorescent lights, 234-240
The Compassionate Consumer, 112
composite packaging materials, 34, 38

composting, 51, 190-191
compressed natural gas, 58-59
concentrates, 36, 37
Concern, Inc., 312-313
Conservation Foundation, 313
Conservatree, 279
The Conserve Group, 209
consumer boycotts, 294-297
Consumer Pesticide Project, 313
Consumer Reports, 67, 159, 165
convenience foods, 147-148
corrugated cardboard, 47
cosmetics, 89, 102-114, 110
    chemicals in, 104-106
cotton, 90-92
Council of Better Business Bureaus, 293
Council on Economic Priorities, 313
Cousteau Society, 313-314
The Crafts Center, 283
cruelty-free products, 89, 108-114
Cultural Survival, 283, 314

The Daily Planet, 276
dairy products, 136-137
Dano Enterprises, 191
de-icing salts, 31
Defenders of Wildlife, 314
DejaShoes, 97
dental products, 110
Desert Fishes Council, 314
Desert Tortoise Council, 314
detergents, 159-160
Deva Lifeware, 277-278
Dial Corp., 167, 168
diapers, 98-102
    costs of, 101
dioxins, 153-154, 156-157
Directory of Environmental Travel & Volunteer Activities, 266
dishwashers, 213-214
dishwashing liquids, 169-170
disinfectants, 170
disposable diapers, 29, 98-102
disposables, 28-29, 98-102
Do Dreams Music Company, 275
dolphin-safe tuna, 143-145
Don't Be Cruel, Inc., 114
drain cleaners, 170, 171, 246
drinking water, 31-32, 129-133
dryers, 165
Ducks Unlimited, 314
DuPont, 94
Earth Care Inc., 95
Earth's Best Baby Food, 150
EarthAlert, 276
Earth Care Paper, Inc., 280
Earthen Joys, 114, 157
Earth First!, 314-315
Earth Island Institute, 315
Earthlings, 95-96
EarthSave, 315
Earthwatch, 266-267
The East Bay Depot, 13
eco-labeling, 51-52, 255, 259
Eco-Safe Products, Inc., 201
Eco Bella, 114
Ecolobag, 191
EcoSource, 149
Ecosport, 96
ecotourism, 261-272
The Ecotourism Society, 266
Ecover, 160, 165
EcoWorks, 240
edible packaging, 47
educational campaigns, 291
Educational Insights, 276
EJW Products, 280
electric cars, 61-64
Electric Engineering, 281
Elsa Wild Animal Appeal, 315
Victor Emanuel Nature Tours, 269
The Emperor Wears No Clothes, 93
energy-efficient appliances, 165-166, 204-227
energy-efficient lighting, 233-240
energy-saving products, 228-232
Energy Answers Corp., 100
energy conservation, 15-16, 18, 22-23, 203-240
"Energy Conserving II" motor oil, 83
The Energy Store, 232
Environmental Action, Inc., 315-316
Environmental Action Coalition, 315
Environmental Action Foundation, 289
Environmental and Energy Study Institute, 316
Environmental Awareness Products, 278
environmental claims, 293
Environmental Defense Fund, 316
Environmental Hazards Management Institute, 316
environmental investments, 298-301
environmental labeling, 51-52, 255, 259
Environmental Law Institute, 316

Environmental Protection Agency, 21, 23, 27, 28, 31, 33, 40, 82, 87, 118, 176, 177, 183, 190, 196, 197, 256
Environmental Safety, 316
Environmental Traveling Companions, 271
Equal Exchange, 283
Esprit, 96
Estee Lauder, 103
ethanol, 58
Evergreen Alliance, 193
Exxel Container, 103
Exxon *Valdez*, 80

fabrics, 90-98
fabric softeners, 164
Farm Verified Organics, 124
fast food, 141-142
Faultless Starch/Bon Ami Co., 167
Federal Trade Commission, 152
fire extinguishers, 20
fish, 141, 143-145
Ron Fisher Furniture, 256
flea and tick control, 170
fleas and ticks, 195-202
flexible-fuel engines, 58
floor cleaners, 171
Flush-N-Save, 245
Food and Drug Administration, 31
food and groceries, 115-174
food safety, 118-122
    see also organic foods
forests, 192-194
    see also rain forests
The Forest Trust, 317
Fox Fibre, 95
fragrances, 110
freezers, 212-213
Freshwater Foundation, 317
Friends of Animals, 317
Friends of the Earth, 317
Friends of the River, 317
Friends of the Sea Otter, 317
Frontier Cooperative Herbs, 201
frozen foods, 137-139
fruits, 145-147
    see also organic foods
fuel-boosting gadgets, 77
fuel-efficient cars, 16, 56-64, 64-68
Fund for Animals, 317-318
furnaces, 224-227
furniture, 253-257
furniture polishes, 171

Gallup Poll, 6
garbage, *see* solid waste
gardening, 199
garden supplies, 175-191
    low-impact, 188-189
gasohol, 58
gasoline, 22-23, 74-78
    brands of, 76-77
    reformulated, 60-61
    see also alternative fuels
Gates Energy Products, 242-243
General Electric Co., 240, 243
General Motors, 56-57, 58, 63, 64
gifts, 273-284
Gillis, Jack, 67
glass cleaners, 171
glass containers, 29, 33, 48
The Glidden Company, 252
Global ReLeaf, 294
Global Tomorrow Coalition, 318
global warming, 16-18, 87, 119
Goldbeck, David, 191
grass, 185-188
Grass Roots the Organic Way, 318
Great Bear Foundation, 318
Greater Yellowstone Coalition, 318
Great Swamp Research Institute, 318
"green" cotton, 91-92, 95-98
Green At Work, 305
Green Cross, *see* Scientific Certification Systems
Greenhouse Crisis Foundation, 318
greenhouse effect, *see* global warming
Greenpeace, 289, 292, 319
green products, definition of, 8
Green Seal, 51-52
greenwashing, 293
GTE Sylvania Lighting Div., 240
Guatemalan Refugee Crafts Project, 283

hair care products, 104, 110-111
hamburgers, 141-142
    see also beef
Hand In Hand, 151
Harding Energy Systems, 243
hardware stores, 230-231
Louis Harris poll, 175
Heart of the Dolphins, 275
heaters, *see* furnaces
heat pumps, 221-223
heavy metals, 34
H. J. Heinz Co., 143
hemp, 93

Hershkowitz, Allen, 99, 100
Honda Motor Co., 61
household cleaners, *see* cleaning products, laundry detergents
Household Hazardous Waste Project, 158, 162, 319
house paint, 247-252
How Much Is Enough?, 11
Hubbard Scientific, 276
Humane Alternative Products, 114
Human Ecology Action League, 319
Human Environment Center, 319
Humane Society of the United States, 108-109, 319

Infact, 319-320
Inform, 320
insect growth regulators, 201
insecticides, 195-202
Institute for Earth Education, 320
Institute for Local Self-Reliance, 320
integrated pest management, 179-189
Interior Concerns newsletter, 257
International Alliance for Sustainable Agriculture, 122, 143, 320
International Council for Bird Preservation, 320
International Expeditions, Inc., 268
International Fund for Animal Welfare, 320-321
International Oceanographic Foundation, 321
International Primate Protection League, 321
International Society of Arboriculture, 321
International Society of Tropical Foresters, 321
investments, 298-301

Jade Mountain, 232, 280
jobs, environmental, 305
Johnny's Selected Seeds, 188
S. C. Johnson & Son, 105
Journeys, 268
Juice Bowl Products, Inc., 128
juices, 134

Robin Kay, 97
Kingsley-Bate Ltd., 256
kitchen wraps and bags, 151-153
kitty litter, 201
The Knoll Group, 254, 255, 257

labeling, environmental, 51-52, 255, 259
landfills, 26-29, 42, 79, 98
laundry detergents, 159-160, 162-164, 166-168, 172
lawns, 185-188
League of Conservation Voters, 291, 321-322
League of Women Voters of the United States, 322
letter-writing campaigns, 290-291
life-cycle costing, 207
lighting, 233-240
Little Red's World, 114
livestock, 93
Livos PlantChemistry, 252
Lost Valley Center, 271
low-impact garden supplies, 188-189
lunches, 149

Maho Bay Camps, Inc., 271
Making Polluters Pay, 289
Marine Mammal Stranding Center, 322
marketing claims, 293
Marketplace, 278, 283
materials, 90-98
Mayan Hands, 283
Mazda, 63
meat, poultry, and fish, 139-145
meat, barbecuing, 147
mechanics, automobile, 73
Melitta U.S.A., 157
metal polishes, 172
methane, 17
methanol, 57
microwave ovens, 138-139
milk, 136-137
Miller Paint, 252
Mitsubishi Motors Corp., 61
mold and mildew cleaners, 172
Monitor Consortium of Conservation and Animal Welfare Organizations, 322
Morrills' New Directions, 201
mothballs, 172
motor oil, 71, 80-82, 83
 disposing of, 81-82
 recycling, 50
Murco Wall Products, 252
Music for Little People, 279
mutual funds, socially responsible, 298-301

Nader, Ralph, 75

nail care products, 111
National Academy of Sciences, 118, 119, 175
National Arbor Day Foundation, 194, 322
National Association of Diaper Services, 102
National Association of Towns and Townships, 292-293
National Audubon Society, 263, 266, 270, 322-323
National Boycott News, 295, 297
National Clean Air Coalition, 323
National Coalition Against the Misuse of Pesticides, 195, 198, 323
National Coalition for Marine Conservation, 323
National Gardening Association, 175, 176
National Geographic Society, 276, 323
National Institute for Urban Wildlife, 323
National Organization for the Reform of Marijuana Laws, 93
National Parks and Conservation Association, 272, 323-324
National Park Service, 264, 272
National Recycling Coalition, 325
National Toxics Campaign, 325
National Water Center, 325
National Wildlife Federation, 274, 281, 290, 291, 325
Natural Brew, 157
The Natural Choice, 275, 281
natural gas for automobiles, 58-59
natural pest control, 182, 184
Natural Resources Defense Council, 99, 119, 325
Natural World, Inc., 114
The Nature Company, 276-277
The Nature Conservancy, 326
Nature Expeditions International, 268
Necessary Trading Co., 201
The New Green Christmas, 193
Nissan Motor Co., 63
nitrous oxide, 17
North American Lake Management Society, 326
nylon, 94

octane, 74-76
Odara, 283-284
oil, *see* motor oil
oil companies, 75
oil filters, 55, 71
Organic Crop Improvement Association, 124
organic farming, 122-126
organic foods, 18, 123-126, 142-143, 150
Organic Foods Production Association of North America, 125
Organic Growers and Buyers Association, 125
organic meat, 142-143
Osram, 240
oven cleaners, 172
overpackaging, 33-34, 36-37
Overseas Adventure Travel, 264
O Wear, 96
ozone depletion, 18-20, 34, 104

Pacific Whale Foundation, 326
packaging, 33-43, 89, 102, 116, 144, 145, 151, 160-161
  best and worst, 35-39
  edible, 47
paint, 247-252
Panasonic, 240
Panti Medicine Trail, 25
paper and paper products, 29, 153-157, 279-280
paper recycling, 48-50
Patagonia, 96
Pennsylvania Resources Council, 326
People for the Ethical Treatment of Animals, 109, 113, 326
The Peregrine Fund, 326-327
pest control, 170, 182, 184
Pesticide Action Network, 327
pesticide labels, 178-179
pesticides, 118-126, 174, 176-189
Petco Animal Supplies, 202
Pet Guard, 201
pet supplies, 195-202
Philips Lighting, 240
phosphates, 159-160
plastics, 27, 29, 33, 39-43, 128, 135-137, 160-161
  recycling, 42-43, 50-51
  types of, 41
Plow & Hearth, 257
political campaigns, 291
poultry, 139-140, 145
powdered mixes, 135
Practical Home Energy Savings, 229

Pristine Products, 201
Procter & Gamble, 103, 105, 167-168
produce, 145-147
see also organic foods
Programme for Belieze, 281
Public Citizen, 75, 327
Public Voice for Food and Health Policy, 327
Pueblo to People, 284
Pure Podunk, 97

Rachel Carson Council, 327
Radio Shack, 243
Rails-to-Trails Conservancy, 272
Rainforest Action Movement, 327
Rainforest Action Network, 141-142, 258, 259, 260, 297, 327-328
Rainforest Alliance, 255, 259, 328
rain forest beef, 141-142
Rainforest In Your Kitchen, 131
rain forests, 23-226, 281
razors, 99, 111
re-refined motor oil, 81
Real Goods Trading Co., 232, 277, 280
rechargable batteries, 12, 241-243
Recipes from an Ecological Kitchen, 141
Recursos, 270
recycled paper products, 155
recycling, 12-13, 34, 42-43, 44-51, 128, 292, 302-303
reducing purchases, 10-12
refillables, 36, 37
reformulated gasoline, 60-61
refrigerators and freezers, 208-213
refurbishing, 13
Renew America, 328
repairing, 13
reporting polluters, 32, 288-289
Resources Conservation, Inc., 246
Resources for the Future, 328
reusing, 12, 13
Rifkin, Jeremy, 140, 143
Ringer Corp., 188, 201
Rocky Mountain Humane Society, 92
Rocky Mountain Institute, 229, 328
John Rossi Company, 280
Rubbermaid, Inc., 149
rubber recycling, 51

Safaricentre International, 268
Sanyo Energy (USA) Corp., 243
Sass, Lorna J., 141
Save the Redwoods League, 328
school lunches, 149
Scientific Certification Systems, 51-52, 90, 152, 161, 255
seafood, 141, 143-145
Sea Shepherd Conservation Society, 328-329
Selfhelp Crafts of the World, 284
Sergeant's Pet Products, 103
SERRV, 284
Seventh Generation, 91-92, 157, 232, 246, 278, 280
shampoos, 110
shaving products, 111
Sierra Club, 290, 329
Silent Spring, 177
Silk for Life Project, 284
skin care products, 111, 112
Skinner, Nancy, 249
The Smart Kitchen, 191
Smart Wood, 255, 259
Smith & Hawken, 257
Smithsonian National Associate Program, 270
snacks, 147-148
Social Investment Forum, 301, 305
socially responsible investments, 298-301
Society for Ecological Restoration, 270
sodas, 135-136
Soil and Water Conservation Society of America, 329
solar-powered products, 280-281
Solar Car Corp., 64
Solar Electric Engineering, Inc., 64
solid waste, 26-29, 39-43
see also packaging
spark plugs, 71
The Sprout House, 188
starch, 164
StarKist, 144
steel containers, 51
J. P. Stevens, 92
Levi Strauss & Co., 96
Student Conservation Association, 270, 329
The Student Environmental Action Guide, 295
Student Pugwash, 329
Summit Furniture, 257
sun care products, 112
Sunrise Lane Products, 114
Sun Watt Corp., 243, 281
supermarkets, 116-117

Sustainable Transportation Repair Society, 73
Suzuki Motor Co., 57
Swissgold, 157
synthetic materials, 94-95

tampon applicators, 99
Tamu Safaris, 268
Tarlton Foundation, 281
Teitel, Martin, 131
Thai International Display, 284
"three R's, " 10-13
tires, 51, 55, 71, 87
toilet cleaners, 174
toiletries, 112
Tom's of Maine, 102, 107
toothbrushes, 112
Topline Instant Drain Opener, 246
Toyota Motor Co., 61
toys, 273-284
Trade Wind, 284
trash bags, 151-153
travel, 261-272
Travel Link, 262, 266
Tread Lightly Limited, 268-269
TreePeople, 194, 330
trees and forests, 18, 192-194, 281, 293-294
Trees for Life, 194, 330
Trust for Public Land, 330
tuna, 143-145
TV dinners, 137-139

U.S. Forest Service, 272
U.S. Public Interest Research Group, 121, 330
Ultracel, 253
underground storage tanks, 30
Union of Concerned Scientists, 16, 330
United Nations, 23-24
University Research Expeditions Program, 270
Urban Ore, 13
used cars, 67
Use It Again, Seattle!, 13
Utica, 97

Valdez Principles, 304-305
vanpooling, 87-88
vegetables, 145-147
see also organic foods
Volvo, 61
Voyagers International, 269
VPSI, 88

Walnut Acres, 114
Izaak Walton League of America, 321
washing machines, 165, 165, 214-215
Washington Toxics Coalition, 195
Wastebusters, 13
water, bottled, 31-32
water conservation, 32, 244-246
water filters, 132-133
water heaters, 165, 215-217
water pollution, 29-32, 118, 119-120
water softeners, 163
Wearable Arts, 278
Wearable Integrity, 97-98
Wearables Marketing, 98
The Whale Center, 330-331
whales, 281
Wicahpi Vision, 284
Wilderness Society, 21, 331
Wilderness Travel, 269
Wildland Adventures, 267
Wildlife Conservation International, 331
Wildlife Information Center, 331
Wildlife Society, 331
Wildlife Tourism Impact Project, 266
Windstar Foundation, 331
wood and wood products,26, 253-260
The Wood Users Guide, 260
wool, 92-93
workplace environmentalism, 302-305
World Environment Center, 331-332
World Resources Institute, 332
Worldwatch Institute, 11, 332
WorldWIDE, 332
World Wildlife Fund, 332

The Xerces Society, 332

yard wastes, 51
Yesterday's News, 201

Zegrahm Expeditions, 269
Zoo Doo, 197
zoos, 271-272